Laboratory Manual

Chemistry in Context

Applying Chemistry to Society

Eighth Edition

Edited by

Jennifer A. Tripp
San Francisco State University

and

Lallie McKenzie
Chem11, LLC

ACS
Chemistry for Life®

CHEMISTRY IN CONTEXT LABORATORY MANUAL: APPLYING CHEMISTRY TO SOCIETY,
EIGHTH EDITION

Some ancillaries, including electronic and print components, may not be available to customers outside the United
States.

This book is printed on acid-free paper.

1 2 3 4 5 6 7 8 9 0 QVS/QVS 1 0 9 8 7 6 5 4

ISBN 978—0—07—351812—1
MHID 0—07—351812—3

Senior Vice President, Products & Markets: *Kurt L. Strand*
Vice President, General Manager, Products & Markets: *Marty Lange*
Vice President, Content Production & Technology Services: *Kimberly Meriwether David*
Managing Director: *Thomas Timp*
Executive Brand Manager: *David Spurgeon, Ph.D.*
Director of Development: *Rose Koos*
Development Editor: *Jodi Rhomberg*
Director of Digital Content: *Andrea M. Pellerito, Ph.D.*
Marketing Manager: *Heather Wagner*
Director, Content Production: *Terri Schiesl*
Content Project Manager: *April R. Southwood*
Buyer: *Nichole Birkenholz*
Designer: *Tara McDermott*
Cover Image: © *Lou Paintin/Getty Images/RF*
Lead Content Licensing Specialist: *Carrie K. Burger*
Compositor: *Aptara, Inc.*
Typeface: *10/14 Times New Roman*
Printer: *Quad/Graphics*

Some of the laboratory experiments included in this text may be hazardous if materials are handled improperly or if
procedures are conducted incorrectly. Safety precautions are necessary when you are working with chemicals, glass
test tubes, hot water baths, sharp instruments, and the like, or for any procedures that generally require caution.
Your school may have set regulations regarding safety procedures that your instructor will explain to you. Should
you have any problems with materials or procedures, please ask your instructor for help.

The Internet addresses listed in the text were accurate at the time of publication. The inclusion of a website does not indicate
an endorsement by the authors or McGraw-Hill Education, and McGraw-Hill Education does not guarantee the accuracy of
the information presented at these sites.

www.mhhe.com

Table of Contents

Preface to Instructors .... *v*

To the Student .... *vi*

Green Chemistry .... *vii*

The Science Writing Heuristic .... *ix*

Safety in the Laboratory 1

Laboratory Methods 3

Investigation 1 – Preparation and Properties of Gases in Air 21

Investigation 2 – Extracting Limonene With Liquid CO_2 27

Investigation 3 – Chromatographic Study of Dyes and Inks 35

Investigation 4 – Graphing the Mass of Air and the Temperature of Water 39

Investigation 5 – What Protects Us from Ultraviolet Light? 47

Investigation 6 – Color and Light 53

Investigation 7 – Testing Refrigerant Gases 59

Investigation 8 – Molecular Models, Bonds, and Shapes 67

Investigation 9 – Measuring Molecular and Molar Mass 73

Investigation 10 – Verifying Molar Ratios in Chemical Reactions 81

Investigation 11 – Hot Stuff: An Energy Conservation Problem 87

Investigation 12 – Comparing the Energy Content of Fuels 89

Investigation 13 – Preparation and Properties of Biodiesel 95

Investigation 14 – Detecting Ions in Solution 101

Investigation 15 – Analysis of Vinegar 107

Investigation 16 – Measuring Water Hardness 113

Investigation 17 – Measuring Chloride in Water Samples 119

Investigation 18 – Analyzing Bottled Water 125

Investigation 19 – Reactions of Acids with Common Substances 131

Investigation 20 – Characterizing Acidic and Basic Materials 137

Investigation 21 – Acid Rain 141

Investigation 22 – Investigating Solubility 145

Investigation 23 – Measuring Radon in Air 147

Investigation 24 – Exploring Electrochemistry 153

Investigation 25 – Polymer Synthesis and Properties 161

Investigation 26 – Identifying Common Plastics 169

Investigation 27 – Identifying Analgesic Drugs by TLC 175

Investigation 28 – Synthesizing Aspirin 179

Investigation 29 – Drugs in the Environment 187

Investigation 30 – Measuring Fat in Potato Chips and Hot Dogs 191

Investigation 31 – Measuring the Sugar Content of Beverages 197

Investigation 32 – Measuring Salt in Food 201

Investigation 33 – Measuring Vitamin C in Juice and Tablets 205

Investigation 34 – Isolating DNA from Plant and Animal Cells 209

Glossary.. 213

Preface to Instructors

This laboratory manual accompanies the seventh edition of *Chemistry in Context: Applying Chemistry to Society*. Both the text and this laboratory manual are designed for college students majoring in disciplines outside of the natural and physical sciences. *Chemistry in Context* examines a set of scientific and technological topics with broader societal implications; this manual provides laboratory experiences that are relevant to these topics. As authors and instructors, we believe that laboratory work ought to be an integral part of chemistry courses. Hands-on experience with investigations and data collection are crucial to an understanding of the scientific method and to the role that science plays in addressing societal issues.

The investigations in this laboratory are relatively straightforward with easy-to-follow instructions. Some are adaptations of traditional investigations; others are quite novel. Most of the investigations use small-scale equipment. The investigations require relatively little mastery of traditional laboratory techniques; thus, maximum student time can be devoted to explorations and acquiring data. Some of the investigations can be completed in an hour or less, while others take longer.

The changes for the eighth edition have been extensive, with a major revision of every lab investigation to utilize the Science Writing Heuristic (SWH). We encourage you and your students to read the introduction to SWH in this volume. With SWH, the students ask questions that can be answered experimentally, and then interpret and analyze their data and make claims about their results. Students are encouraged to write about the investigation and devise their own data tables and charts, and therefore the data sheets have been eliminated from the lab manual. Nevertheless, we realize that many instructors find the data sheets useful and we have made them available online along with the instructor's notes. They can be modified to suit the course and printed as desired.

Two brand-new investigations have been added to this edition: an extraction of limonene from orange peel using liquid carbon dioxide, which nicely highlights several of the key ideas in green chemistry; and a measurement of caffeine in water samples that encourages the students to think about the consequences of pharmaceutical runoff. Several other investigations that were dropped from previous editions have returned. Michael Samide and Todd Hopkins of Butler University were instrumental in developing the caffeine lab, and we thank them for their contribution to this volume. We also thank Michael Cann, who wrote the introduction to green chemistry, and Michael Mury, who advised us on the Science Writing Heuristic and assisted in editing.

This collection of investigations is the result of a collaborative effort involving many people. Major contributions to earlier editions came from Catherine Middlecamp, Norbert Pienta, Truman Schwartz, Robert Silberman, Conrad Stanitski, Gail Steehler, and Wilmer Stratton.

All of the investigations in this lab manual have been used by *Chemistry in Context* authors, past and present, in our own classes. They are versions that we have found to fit our needs. But each course is different, and instructors have differing styles. We invite users to contact us with suggestions for modifications of these investigations and/or for exchange of ideas about possible new investigations.

To the Student

The investigations in this laboratory manual have been carefully chosen and designed to reflect and amplify the topics in *Chemistry in Context*. Our goal in writing this manual was not to train you as a future chemist, but rather to illustrate what chemists do and how they do it. We also hope that you will appreciate that there is no great mystery to doing chemistry in the laboratory. You can obtain a great deal of information about the world around you with simple chemical equipment and straightforward procedures.

Another goal is to give you an opportunity to discover how chemists solve problems and to solve some yourself. After all, chemistry is a science of investigation and most chemists spend part of their time in a laboratory. You will use the laboratory to try out new ideas, investigate the properties of materials and compounds, synthesize compounds, analyze materials, and, in general, solve problems. Investigations are the way that scientists answer questions.

Thus, you will find that each of the investigations in this lab manual begins by posing several questions. You will develop hypotheses and procedures for answering the questions. The investigations in this laboratory manual are in the format of the Science Writing Heuristic, a method that encourages you to mindfully develop questions and answer them through investigation. We encourage you to read the short introduction to the Science Writing Heuristic so that you can get the most out of the laboratory component of this course.

You will find that most of these investigations use simple equipment and that you can easily learn the necessary techniques. In general, you will be working with a partner, and in some cases the whole class will work collaboratively to collect data and answer a scientific question. Your instructor may ask you to keep a laboratory notebook or to fill in data sheets with your observations and results. In either case, it is important to keep good records about what you do in the laboratory.

We have been careful, where possible, to develop investigations that use minimally toxic reagents, and to use them in small quantities. You will be encouraged to consider the investigations in the context of **green chemistry**, a way of doing chemistry that seeks to protect the health of humans, wildlife, and the wider environment. Green chemistry is part of a greater movement toward *sustainability*, ideas of which you may have encountered in your other courses. We hope that the investigations in this book, together with ideas presented in the textbook, will enhance your understanding of some of the ways that chemists are helping to solve global problems.

Although you may never again work in a chemistry laboratory, it is our hope that after this laboratory course you will understand why chemists find laboratory work so interesting and compelling.

Green Chemistry

Ask any group of chemists what the world of chemistry has contributed to society and they will come up with a long list. The list might include: pharmaceutical drugs that help us live longer healthier lives and ease our pain and suffering; polymers that make possible dashboards, skateboards, snowboards, and circuit boards; pesticides and fertilizers that without which we would have an even harder time of feeding the 7.1 billion hungry mouths on this planet; flavors and fragrances that we take for granted; cosmetics and personal care products that we use every day; batteries that power our flashlights, toys, cell phones, laptops and start (and even power some of) our cars; and the list goes on and on.

Yet when non-chemists (such as yourselves) are asked what comes to mind when they hear the word "chemical," reciting these wonders of the modern world are generally not how they respond. More often they cite the toxic waste dumps that are down the block or across town from where they live, or depletion of the ozone layer by CFCs, or decimation of the bald eagle population by DDT, or other unpleasant effects of the chemical industry. Chemists, for too long, have not paid enough attention to the environmental consequences of the products they produce nor the processes by which these products are made.

HOWEVER, there is a new paradigm "in town" and it is called **green chemistry**, chemistry with the environment in mind. Chapter 0 of the textbook that accompanies this lab manual, titled Chemistry for a Sustainable Future, introduces you to green chemistry and sustainability. As you can see from Chapter 0 and the key ideas in green chemistry, on the inside cover of this lab manual, green chemistry seeks to reduce the ecological footprint of chemical products and the processes by which they are made. Many of the investigations in this manual have been designed with green chemistry in mind. For example, Investigation 2 involves extracting the essential oil from orange peels using liquid carbon dioxide, a benign solvent that can replace toxic and flammable hydrocarbon solvents. Investigation 7 allows you to compare the properties of refrigerant gases and analyze how CFCs became popular and how alternatives are evaluated. In Investigation 13 you will synthesize biodiesel from vegetable oil, which offers a renewable alternative to petroleum-based diesel fuel used by cars and trucks. In Investigation 28 you can synthesize aspirin, and you might use energy-efficient microwave heating to promote the reaction. Many more investigations in this lab manual use and discuss green chemistry, and we encourage you to think about the key ideas as you do the investigations.

Green chemistry, along with other green technologies, is an integral part of our journey to find a path to sustainable development. As is indicated in Chapter 0, humanity's combined ecological footprint exceeds the carrying capacity of the Earth. In short, we are depleting the natural capital of the Earth (her renewable resources) and producing waste faster than the Earth is able to convert our waste back to natural capital. Green technologies not only offer the possibility of reducing the amount of resources that we consume, but even better, they offer an opportunity to convert our waste back into natural capital. That is, technology becomes part of the solution, not part of the problem. We hope, that by learning about green chemistry and sustainability, this inspires you to develop ways, in both your personal and (future) professional life, that you can contribute to sustainable development.

Michael C. Cann
University of Scranton

The Science Writing Heuristic

If you ask your classmates and even your professors for the definition of inquiry, you will probably get as many unique definitions as the people you ask. Inquiry definitely takes on different meanings for different people. In this lab manual we focus on inquiry as the means for performing science. In order to have you think through the processes of science we use headings and structure based on the Science Writing Heuristic (SWH).

The Science Writing Heuristic is a well-developed approach to guided inquiry experiences that is designed to encourage construction of conceptual knowledge. It is also based upon relationships among questions, evidence, and claims.

The traditional Science Writing Heuristic has the following categories for students to process:

1. Beginning Questions—What are my questions?

2. Tests—What did I do?

3. Observations—What did I see?

4. Claims—What can I claim?

5. Evidence—How do I know? Why am I making these claims?

6. Reading—How do my ideas compare with other ideas?

7. Reflection—How have my ideas changed?

8. Writing—What is the best explanation that clarifies what I have learned?

Based on the work of Dr. Angela Powers and her author team for the textbook *Chemistry in the Community*, we have adapted these categories for you into the headings below that you will see in each investigation.

Asking Questions

Scientific investigations usually begin with a question to be answered through data gathering and experimentation. Sometimes this question will be provided, while other times you will be asked to develop the question with your laboratory partners or classmates.

Preparing to Investigate

Before beginning experiments, it is important to clearly outline a procedure for gathering evidence that includes identifying the data to be collected and the steps to be followed. In some cases a complete or partial procedure will be included in the investigation, but many times you will devise part or all of the procedure with your laboratory partners or classmates.

Whether the procedure is provided or devised, you will need to study it completely before beginning. You will also need to create a system—usually a data table—for recording the observations and measurements you will make during the investigation.

Making Predictions

In some investigations, you will predict what you think will happen as you gather evidence. These predictions should be based on your prior experience and will not be evaluated for correctness, but you may be asked to reflect upon them after the investigation.

Gathering Evidence

Gathering Evidence is the core of the investigation. It contains directions, steps, or guidance for collecting data and observations.

Analyzing Evidence

The evidence gathered in some investigations requires further processing before it is useful in answering questions. Guidance is often provided to facilitate calculations and other analysis.

Interpreting Evidence

The next step after analyzing evidence is to ask "What does the evidence mean?" Answering this question allows you to propose explanations for scientific phenomena. Questions within this section are designed to help you think about implications of the evidence and connect it to the purpose of the investigation.

Making Claims

Once data have been analyzed and interpreted, an answer to the initial question can be proposed. This answer often comes in the form of a scientific claim. Such claims must be supported by evidence from the investigation.

Reflecting on the Investigation

The final task in most investigations is to reflect on what was done, think about how your understanding has developed, and apply what was determined to other situations.

We hope that these categories will assist you as you work through these investigations as well as helping you to think like a scientist!

Michael Mury
American Chemical Society

Source: Greenbowe, T.J. & Hand, B. *Introduction to the Science Writing Heuristic*, in Pienta, N.J., Cooper, M.M., and Greenbowe, T.J., eds. *Chemists' Guide to Effective Teaching*, Prentice Hall 2005

Safety in the Laboratory

Good laboratory practice requires that you take some simple safety precautions whenever you work in a chemistry laboratory. The popular notion that a chemistry laboratory is a dangerous place, filled with unknown disasters waiting to occur, is simply untrue for most situations and is certainly incorrect for the activities in this laboratory manual. Nevertheless, all chemistry laboratories have some hazards associated with chemical spills, careless handling of flammable substances, and broken glassware. With these in mind, here are some basic rules to follow when working in a chemistry laboratory.

1. **Protect your eyes with approved goggles or glasses.** This rule is essential and will be rigidly enforced. Failure to wear approved eyewear will result in dismissal from the laboratory. Chemical splashes can harm your eyes, and even dilute solutions of many chemicals can cause serious eye damage. You must wear eye protection even when you are not working directly with chemicals. Someone near you may have an accident and something may splash or fly in your direction.

2. **Exercise special care when using flammable substances.** Tie back long hair and avoid wearing loose sleeves. No open flames should be used anywhere in the laboratory, unless the experiment specifically calls for the use of a burner. Even a hot object can sometimes cause flammable vapors to ignite, so it is important for you to know which liquids are flammable.

3. **Never eat, drink, or smoke in the laboratory.** You may inadvertently ingest or inhale hazardous chemicals. It is also a good idea to wash your hands before exiting the laboratory.

4. **Never perform unauthorized experiments.** Some simple chemicals can form explosive or toxic products when mixed in unintended or inappropriate ways.

5. **Never work in a laboratory without proper supervision.** One of your best safety precautions is to have a knowledgeable person present who can spot potential hazards and handle an emergency should it arise. Notify your instructor immediately in case of any spill or injury, no matter how small. If you have any questions or concerns about safety, please address those with your instructor before proceeding with an experiment.

6. **Handle glassware carefully.** Glassware can break and cause nasty cuts. Report broken glassware to your instructor and dispose of it properly.

7. **Learn the location of safety equipment in your lab, including fire extinguishers, fire blanket, first-aid kit, eyewashes, and safety showers.** Be sure you know how and when they are to be used.

 STOP! Important safety information in each experiment is marked with this icon. Pay close attention to these cautionary notes and any additional safety information given by your instructor. Develop a habit of being safety-conscious whenever you are in the laboratory.

Notes

Laboratory Methods

In this laboratory course, you will be exploring the physical and chemical properties of matter. Because we cannot directly see molecules, we will be using a number of more indirect techniques, some of which may be unfamiliar to you. This section will introduce and explain the most common methods and procedures that you will use in the chemistry laboratory.

Measuring physical properties

Much of what you will be doing in the laboratory involves taking measurements of a number of properties such as mass, volume, and temperature. Here, we will go over a few basic techniques for measuring these different properties. Keep in mind that actual laboratory equipment can vary and you should pay close attention to the instructions given by your instructor for the safe use and operation of the equipment available in your laboratory.

Measuring mass

The **mass** of a substance is a measure of how much matter it contains. Typically, mass is measured using an analytical balance such as that pictured in Figure 0.1. The balance has a plate or pan to hold your sample, a display to read the mass, and often a clear box or dome that goes over the sample to prevent air currents from disrupting the measurement. In addition, there will be some buttons near the display.

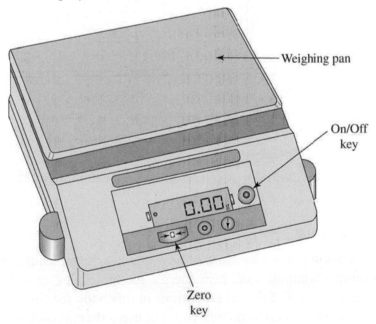

Figure 0.1. Analytical balance

Investigations in this laboratory will require a balance with an accuracy to at least 1/100th of a gram (0.01 g, or 10 mg). To measure the mass of a sample, it is important to make sure that your balance reads exactly zero before making a measurement. Make sure that the balance pan is empty and press the "tare" or "zero" button. Your instructor will show you how to do this with your specific balance. Only after the display reads exactly zero should you put your sample on the pan. Allow the instrument to settle and write down the mass of your sample. Be sure to record the entire mass, even if the last digit is zero.

Measuring volume

Beakers, Erlenmeyer flasks, and other shapes of laboratory glassware have volume markings, but these markings are often not very accurate. The best way to measure the **volume** of a liquid sample is with a graduated cylinder (Figure 0.2). This piece of glassware has evenly spaced markings along the side of the glass cylinder, enabling you to read the volume with high accuracy. To measure the volume of a liquid, place the sample inside the cylinder and place the cylinder on a flat surface, such as the lab bench. Place yourself at eye level with the top of the liquid. You will see that most liquids are not flat on top, but rather have a curved surface called a **meniscus**. You should read the volume at the *bottom* of the meniscus, as shown in Figure 0.2. Read the value and then estimate the position between the two lines. For instance, the volume in Figure 0.2 should be read as 25.7 mL.

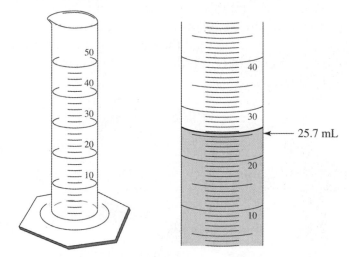

Figure 0.2. The volume of a liquid in a graduated cylinder should be read at the bottom of the meniscus.

Measuring temperature

The **temperature** of a sample is a measure of how quickly the molecules in the sample are moving. When a sample contains more heat energy, the molecules move more quickly and the temperature is higher. Several of the investigations in this book require the accurate measurement of temperature, and to do this you will use a **thermometer**. You will most likely be using an analog thermometer, a thin glass graduated tube filled with alcohol that has been dyed red. To measure the temperature, insert the bulb of the thermometer into the sample, and allow the red alcohol to come to a stable position. Read the temperature using the marked lines, and then estimate the position between the final two lines for additional accuracy.

Measuring temperature is even easier with a digital thermometer. If one is available, you simply insert the probe into the sample and read the temperature off the digital display.

Measuring pressure

Pressure is the amount of force that a substance, often a gas, exerts on its surroundings. Atmospheric pressure is measured with a **barometer**, which traditionally consists of a column of mercury connected to a bulb. The column of mercury moves up and down depending on the ambient air pressure. Atmospheric pressure at sea level is typically 760 mmHg, but can vary with elevation and weather patterns. More modern instruments for measuring air pressure do not involve mercury, a toxic metal. Should you be required to measure atmospheric pressure, your instructor will provide instructions about how do it.

Pressure can differ significantly from atmospheric pressure in enclosed systems such as tires or soda bottles. If you are required to measure the pressure inside a closed system, you will use a pressure gauge, such as that used to check the air pressure in your car or bicycle tires. These come in many varieties (Figure 0.3) and your instructor will demonstrate the use of the ones you have available.

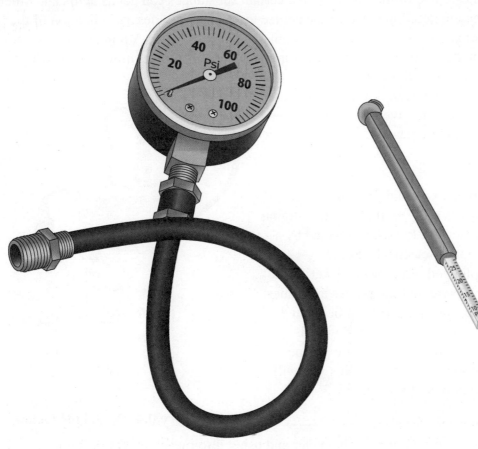

Figure 0.3. Examples of tire pressure gauges.

Additional measuring methods

Some investigations in this book require you to measure other quantities, such as the conductivity, density, or melting point of a substance. Details about these measurements will be given in the individual lab procedures.

Analyzing chemical properties

In contrast to physical properties, chemical properties are those that result from molecular-level change within a substance. Properties such as pH, heat of combustion, and chromatographic affinity are useful for studying chemical reactions and analyzing the chemical composition of a substance. Several instrumental and **wet chemistry** methods will be used in this course.

Measuring pH

Indicator paper

You will learn over the course of the semester that certain substances can act as **acids** and others as **bases**. **pH**, described in detail in Ch. 6 of the course textbook, provides an indication of the presence or absence of acids, as well as their strength and concentration. A number of ways to measure pH have been developed. A simple measurement of pH is to use litmus paper, which turns red with acid and blue with base. More sophisticated pH papers have been developed that use more than one indicator dye to give different color changes over the entire pH range.

Digital pH meter

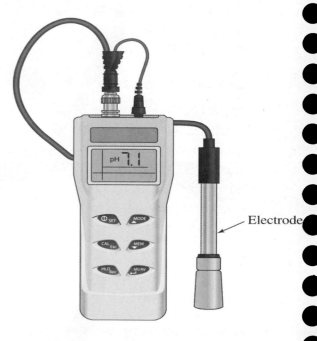

You may also measure pH with a digital pH meter (Figure 0.4). This consists of a small box with a digital display and several buttons or knobs, connected by a cable to a probe called a pH electrode. Some instruments must be plugged into a wall socket, while others run on batteries. Note that the probes are rather fragile and also very expensive, and must be treated with care. The probe should be rinsed and wiped carefully and never allowed to dry out. The pH meter must be calibrated before use, and to do this you will use standard buffer solutions of known pH.

Figure 0.4. Digital pH meter

Follow these instructions for calibrating the meter and using it to measure pH of a sample:

1. Obtain two small beakers containing the standard buffer solution. These usually come in pH 4.00, 7.00 and 10.00. Your instructor will indicate which to use for your investigation.

2. Hold the electrode over a waste container (usually just a beaker) and rinse the electrode thoroughly with pure water from a wash bottle. Blot the end of the electrode *gently* with a soft tissue to remove most of the water.

3. Insert the electrode into the higher pH calibrating solution, stir gently for a few moments, and observe the reading on the meter. Use the "calibrate" knob or buttons to adjust the meter until it displays the correct pH, within 0.1 pH units (i.e. for a pH 7.00 reference solution, the calibrated meter should read between 6.9 and 7.1).

4. Again, rinse the electrode thoroughly with pure water, blot the end gently with a tissue, and then insert the electrode into the second, lower pH calibrating solution. Stir and observe until the reading is steady. Use the "slope" or "temperature" control, as indicated by your instructor, until the display reads close to the correct pH.

5. Repeat the measurements with the two buffer solutions (rinsing and blotting each time the electrode is moved) to ensure that the readings are steady.

6. To measure a sample: Thoroughly rinse and gently blot dry the pH electrode. Insert the electrode into the sample and stir gently. Without touching any knobs or buttons write down the pH indicated on the display. Between samples, wash and blot dry the electrode. When the electrode is not in use, it must stay submerged in water or another solution as indicated by your instructor so that it does not dry out.

Other methods

A third method for measuring pH is titration, which uses a chemical indicator and volumetric analysis to determine the concentration of acid or base in a solution. Titration is described in detail below.

Use of indicators

Many chemical changes are invisible so other ways to observe and study them must be applied. A simple way to determine if a chemical change has taken place is to use an **indicator**. Indicators are chemical substances added to reactions to provide a visible signal, usually a change in color or formation of a solid precipitate, that demonstrates that another change has taken place. In this course you may use indicators that change with changes in pH, concentrations of various anions and cations, and vitamin C. In general, you will add a drop or two of the indicator solution to the reaction or sample, and then observe the appropriate change.

Titration Method of Analysis

The **titration** method of analysis is extremely useful for determining the presence and concentration of reactive molecules or ions in a solution. In a titration, a known volume of the solution containing the substance of interest is measured out, then a solution of a compatible reactant is slowly added until a complete reaction is observed. For instance, if you wish to determine the amount of acid in a solution, you would measure a known quantity of the acid and then slowly add a base until you neutralize the solution. If the concentration of the solution of base is known, then you can calculate the concentration of the acid. To determine the endpoint of a titration, an appropriate indicator is added so that it can be observed.

The titrations performed in this course will be done on a small scale, using a 24-well plastic wellplate (Figure 0.5) for the reactions and plastic transfer pipets to add the solutions to the wells. For colored indicators, you should place the wellplate on a white piece of paper to best observe the color change. The best way to observe formation of a solid precipitate is to place the wellplate on a black countertop or piece of paper. Be sure to clearly label all solutions and pipets so that your reagents do not get mixed up during the titration.

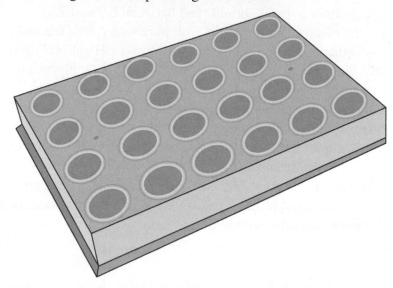

Figure 0.5. The reaction wellplate

Rather than directly measure the volume of the solutions, we will be using plastic pipettes to drop our solutions into the wellplate. Doing this, one assumes that each and every drop from the pipette contains the same volume. The first time you perform a titration, you should practice using the pipet. Fill a plastic pipet with water and try dispensing drops into a well of the wellplate. To successfully perform a titration, you will need to be able to confidently add a known number of drops to the bottom of a well, one drop at a time. Squeeze the bulb gently, while the pipet is held vertically and directly over the center of the well. You may find it helpful to use two hands to steady the pipet. Plastic transfer pipets easily acquire air bubbles in their stems that lead to frustrating partial drops that introduce errors. When you are doing a titration, you should reserve one well for solution waste. During a titration these partial drops can be added to the waste well rather than the titration well. Once you can confidently add a known number of whole drops to a well, proceed to the titration.

It is a good idea before doing a titration to do a trial run. Add 10 drops of your sample and a drop of indicator into a well, and then add your reactant solution dropwise until you observe the indicator change. You may need to stir the reaction with a small stirrer or toothpick for the change to persist. Continue adding a few more drops of the reaction solution to see the extent of the chemical change. Doing this trial run will let you know what to look for when you do your actual titrations.

In doing titrations, your goal is to catch the point where *one drop* of the reactant solution causes the first permanent change in the indicator. You will record the number of drops required to do the titration, and use this along with the known concentration of the reactant solution to calculate the unknown concentration of the sample. Further details about the procedures and the calculations will be given in the investigations that require titration analysis.

Spectroscopy

Spectroscopy is the study of how electromagnetic radiation interacts with matter. The full electromagnetic spectrum includes radio waves, microwaves, infrared, visible, and ultraviolet light, and X-rays (see Ch. 2 in your textbook for more details). Each color and each unique position in the electromagnetic spectrum is identified by its **wavelength**. For example, the visible region contains light with wavelengths in the range of billionths of a meter, and, therefore, the wavelengths are expressed as **nanometers**, nm, (1 nm = 1×10^{-9} m).

In this laboratory course, you will be measuring how molecules absorb ultraviolet and visible light. The instrument used to measure the interaction of light with matter is a **spectrophotometer**. This device is designed to split visible light into its component colors (i.e., different wavelengths) and then allow light of a selected wavelength region to pass through a sample of the material before being studied. An electronic detector measures the amount of light that has been transmitted or absorbed by the sample at each wavelength. All spectrophotometers have essentially the same components but with varying degrees of sophistication. These essential components are: (1) a light source, (2) a device to isolate or resolve particular wavelengths of light, (3) a sample holder, (4) a detector, and (5) a meter or a computer or other device to display the measured transmittance or absorbance of light. Figure 0.6 shows a simple block diagram of spectrophotometer components.

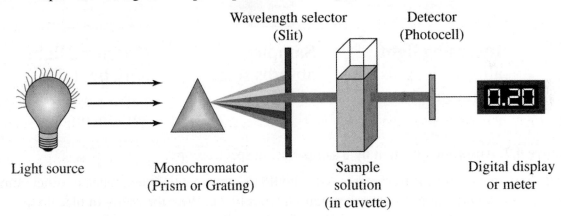

Figure 0.6. Simple diagram of spectrophotometer.

Spectrophotometers are available for most regions of the electromagnetic spectrum. Spectrophotometers for the visible region of the spectrum were developed first and are still the most common. In these instruments, the light source is an ordinary incandescent light bulb; the wavelength selector consists of a diffraction grating and some lenses; a liquid sample is placed in a container resembling a test tube; and the detector is a phototube that converts light intensity into an electrical signal. Different investigations in this book require spectrophotometers that can measure absorbance in the visible and UV regions.

Simple spectrophotometers of the type described here can measure the transmittance of light of only a single wavelength at a time, but it's often useful to measure how a sample interacts with light at a number of different wavelengths (Figure 0.7). Obtaining data over a range of wavelengths requires multiple measurements, each made at a different wavelength. Because light intensity and detector sensitivity vary with wavelength, each measurement must be corrected using a **blank sample**, often pure water.

The meter on the instrument you use will display either **absorbance (A)** or **percent transmittance (%T)**. The percent transmittance focuses our attention on the light that passes unaffected through the sample. Percent transmittance is defined as the ratio of light that passes through a sample to the light that passes through a blank of equal thickness, and multiplied by 100.

$$\%T = \frac{\text{transmittance by sample}}{\text{transmittance by blank}} \times 100\%$$

Absorbance, the other possible spectrophotometer unit, focuses our attention instead on the portion of light that gets absorbed by the sample. Absorbance is proportional to the concentration of the solution. A sample that absorbs no light will have 100% transmission and zero absorbance. Your instructor will let you know which unit to use for the investigations that involve spectroscopy.

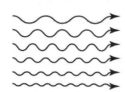 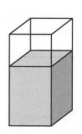 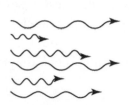

Incoming light: all wavelengths represented

Sample: absorbs some of the light

Outgoing light: only light that wasn't absorbed gets through

Figure 0.7. Absorption of light by a sample using spectroscopy.

When you make your measurements, you should use two identical test tubes or other containers: one containing the sample and one containing your blank. With the blank in place, you will adjust the spectrophotometer to read 100%T or zero absorbance. You will then insert your sample and record data for your sample by simply reading the displayed value. You must re-insert the blank and zero the instrument for each wavelength you measure. Once you measure %T or A over a range of wavelengths, you can plot the data with wavelength along the x-axis and your data on the y-axis. The result is a **spectrum**, a curved line that shows how a particular solution transmits or absorbs light in the visible region of the electromagnetic spectrum. More sophisticated spectrophotometers with computerized detectors may automatically create the spectrum for you.

Spectrophotometers are sensitive and expensive devices, so follow any specific guidelines from your instructor regarding the operation of instruments present in your laboratory.

Chromatography

The term **chromatography** comes from the Greek meaning "to write with color", and indeed the first use of chromatography was to separate and study colored pigments in plants. In its simplest form, a chromatographic set-up consists of an immobilized solid, called the **stationary phase**, over which a gas or liquid, called the **mobile phase**, moves. A variety of chromatographic techniques have been developed, and all chromatographic techniques make use of the fact that components of a mixture injected into the mobile phase can be separated based on their tendency to move along with the mobile phase rather than be attracted to the stationary phase.

A very simple method of chromatography involves using a piece of absorbent paper as the stationary phase and a liquid such as water as the mobile phase. The usual strategy is to place small spots of your sample near one edge of the paper. That edge is placed in the water, and the liquid allowed to move up the paper by capillary action.

A slightly more sophisticated method is **thin-layer chromatography**, or TLC, in which the stationary phase consists of a glass, plastic or aluminum plate coated in a suitable solid substance such as silica or alumina. The mobile phase is a liquid, often a mixture of organic solvents that provide a particular polarity. As above, the sample is spotted on one end of the plate, and that edge immersed in the solvent so that the mobile phase moves up the plate by capillary action, carrying along the components of your sample mixture at different rates to separate them. This is commonly called "developing" the plate.

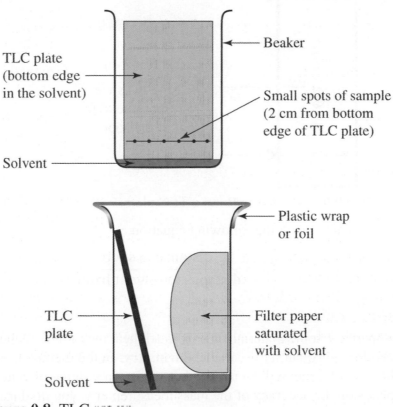

Figure 0.8. TLC set-up

Whether the stationary phase is paper or a TLC plate, it is important to mark your sample spots with pencil, so that you know which samples you have analyzed. Multiple spots may be applied to a single plate or paper, but should be spaced about 1 cm apart. Figure 0.8 shows a typical set-up for chromatography. Depending on the volatility of the solvent used, it may be necessary to cover the beaker with a lid to prevent rapid evaporation. When the solvent front has reached a point about 1 cm from the top of the plate, remove the plate from the solvent, and immediately mark the position of the solvent front with a pencil.

If you are analyzing colored samples, you should be able to directly see them on the plate or paper. If not, some other method will be necessary to make your spots visible. One method is to expose the developed plate to a compound that will react chemically with the spots to make them visible. Another method is to examine the plate under ultraviolet light to see if any of the components are fluorescent. A related technique is to incorporate a fluorescent dye into the plate so that when it is examined under a UV lamp, the plate will glow (fluoresce) everywhere except where the component spots are.

Quantitative analysis of TLC plates is possible by calculating R_f. R_f is defined as the distance traveled by the compound on the plate, divided by the distance traveled by the solvent. It should always be less than 1 since the compounds cannot move further along the plate than the solvent. You calculate R_f by first using a ruler to measure, in mm, the distance from the spot line to the front of the spot, and the distance from the spot line to the solvent front, as shown in Figure 0.9.

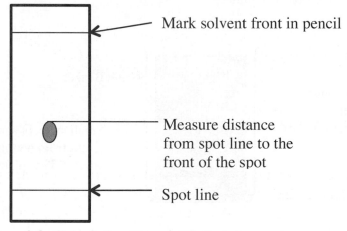

Mark solvent front in pencil

Measure distance from spot line to the front of the spot

Spot line

Figure 0.9. Calculating R_f on a TLC plate.

R_f is then easily calculated using the following equation.

$$R_f = \frac{\text{distance to spot}}{\text{distance to solvent front}}$$

Calorimetry

An important property of some materials is how much heat they release when burned, and the method for measuring this quantity is called **calorimetry**. In the method used in this course, the heat from combustion of a fuel will be used to heat a known volume of water. Accurate calorimetry depends on the accuracy of the mass measurements you do during the investigation, so it's a good idea to review the earlier section on measuring mass before beginning calorimetry. A diagram of the investigation set-up can be seen in Figure 0.10.

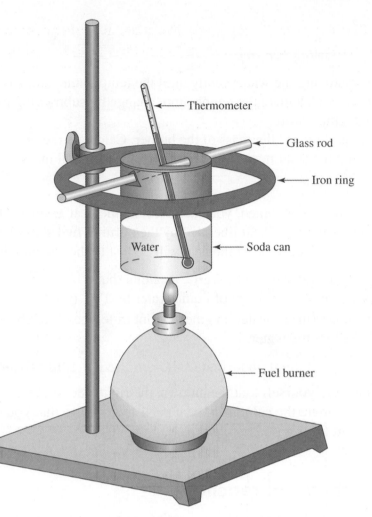

Labels in figure: Thermometer, Glass rod, Iron ring, Water, Soda can, Fuel burner

Figure 0.10. Diagram of can and burner setup

Follow these steps to set up the apparatus and perform a calorimetry investigation.

1. Obtain a dry soda can with the top removed and two holes punched on opposite sides near the top. Measure and record the mass of the empty can.
2. Add approximately 100 mL of water to the can. Measure and record the mass of the can with the water. Calculate the mass of water by subtracting the mass of the can.
3. Slide a glass rod through the holes in the can so that it can be suspended from the ring attached to a ring stand as shown in the diagram.
4. Put a thermometer in the can, stir the water for a few moments, and then measure and record the temperature of the water.
5. Obtain a burner filled with fuel. Measure and record its mass.
6. Place the burner under the can, and adjust the height of the ring so that the bottom of the can is about 2 cm above the top of the wick.
7. Light the burner and observe the flame. If necessary, cautiously adjust the height of the can so that the top of the flame is just below the bottom of the can.
8. Stir the water occasionally and continue heating the water until the temperature has increased by about 20° C; then extinguish the flame.

Quickly do the following two steps:

9. Continue stirring the water gently until the temperature stops rising; then record the temperature. Calculate the temperature change by subtracting the initial temperature from the final temperature.
10. Measure and record the mass of the burner. Calculate the mass of the fuel burned by subtracting the final mass of the burner from the initial mass.

Calculations

Using the data you have obtained, you can calculate the heat absorbed by the water. Theoretically, the amount of heat liberated by the burning fuel should equal the heat absorbed by the water, but in practice, some of the heat will be lost to the surroundings.

The **specific heat** of water is 1.00 cal/g•C, meaning that it takes exactly 1 calorie (cal) of heat to raise the temperature of 1 gram (g) of liquid water by 1° C (see Ch. 5 in the textbook). Therefore, you can use the mass (m) of water and the amount of temperature change (ΔT) to calculate the total heat absorbed by the water.

$$\text{heat absorbed (cal)} = m \times \Delta T \times 1.00 \text{ cal/g•°C}$$

You should convince yourself that the units on the right side of the equation cancel out, leaving only cal. Next, calculate the calories of heat per 1 g of fuel so that you can compare the heat values for different fuels based on their mass.

Running a chemical reaction

In this course you will have several opportunities to transform matter by running a chemical reaction. Of course, most of the methods for analyzing chemical properties described above involve chemical reactions, but when a reaction is run on a large scale with the goal of isolating a product some different techniques are required. Those are described here.

Heating a reaction

Often, one must heat a chemical reaction in order to speed it up. Still other investigations involve heating a substance to measure its physical or chemical properties. You will use several different methods for heating a reaction, each chosen because of its suitability for the particular substance that needs to be heated.

Bunsen burners

Bunsen burners are simply a method to control the flame from burning natural gas. The burner has a diffuser at the top to assist with even heating, and the amount of air entering the burner, and thus the intensity of the flame, can be adjusted by adjusting a valve at the base of the burner.

An important safety note: Bunsen burners should *never* be used to heat flammable substances. Always ensure that no flammable chemicals are in the vicinity of the burner. When you use a Bunsen burner you should tie back your hair and avoid wearing loose sleeves.

Bunsen burners are most often used to heat small amounts of non-flammable solutions, or to perform a flame test, in which a small amount of a solid is placed into the flame to see what color it burns. Heating a test tube over a flame must be done carefully, as shown in Figure 0.11. Heating the tube too rapidly, especially if it is held on an upright position, will cause the hot contents to splash out of the tube. In addition to being quite dangerous, it will require you to start over with a fresh sample.

To perform a flame test, a very small amount of solid or liquid sample is placed on the end of a copper wire, and the wire immersed in the flame. When the flame changes to different colors, it can indicate the presence of certain elements. Further details about flame tests will be given in the investigations that require them.

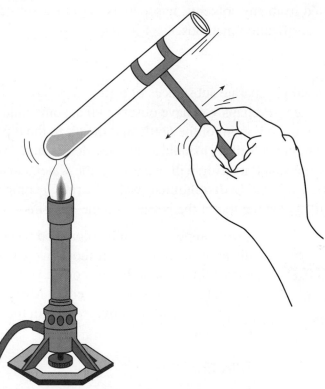

Figure 0.11. Heating a test tube over a Bunsen burner.

Hot plates

Hot plates are an efficient way to heat larger amounts of reagents, and are much safer to use with flammable substances. A good way to heat reactions to a specific temperature is to prepare a hot-water bath, by placing a partially full beaker of water on the hot plate and heating it to a particular temperature as measured by a thermometer. Your test tubes or other reaction containers can then be placed into the water bath to be heated.

Microwave heating

You probably have a microwave oven in your kitchen for cooking and reheating food, but you may not know that they can also be used in the laboratory for heating reactions. Microwaves are a form of electromagnetic radiation that interacts with polar molecules and ions, causing them to rotate rapidly and heat up. Typically, reactions conducted in the microwave take place more quickly and with reduced energy consumption compared to reactions heated in a conventional way. Thus, microwave chemistry is often considered to be a "green" technique. You may have the opportunity to prepare aspirin using microwave heating in this course.

Methods for purification of chemical substances

Measurements often require a pure chemical substance, so knowing some techniques for isolating one compound from a mixture will assist you in your measurements. If you are interested in studying compounds from nature, you often must separate the different components of the plant or food you wish to study. If you perform a chemical reaction, the desired product must be isolated from any solvents, byproducts, or unreacted starting materials. Below, we describe three purification methods.

Extraction

Extraction is used to separate substances of different polarities based on their different solubilities in water. Perhaps you have observed that some kinds of salad dressing separate, with the water going to the bottom of the bottle and the oil to the top. This separation occurs because water and oil are not **miscible**, meaning that they do not mix. Any **hydrophilic** ("water-loving") compounds in the salad dressing will dissolve in the water and sink to the bottom of the bottle with the water, while the **hydrophobic** ("water-fearing") compounds will prefer to dissolve in the oil and will stay at the top of the bottle with the oil. This is exactly the principle of extraction.

Two compounds can be separated if they want to dissolve in two different solvents, and if those two solvents are not miscible. For extractions you do in this course you will use water and a nonpolar solvent such as petroleum ether. More specific details about extraction procedures will be provided within the investigations that require this technique.

Filtration

Filtration is a simple method for separating a solid from a liquid. Perhaps you have used a colander to strain pasta after cooking; the colander acts as a filter to separate the solid pasta from the liquid cooking water. In the chemistry lab, we often use coarse paper as the filter, which is placed into a funnel and the mixture poured through. Any solids get caught in the filter while liquids flow through.

Gravity filtration uses a filter funnel, a cone-shaped glass funnel with a narrow tube extending from the bottom, sometimes with ribs to increase filtering efficiency. Filter paper is folded to fit inside, and the funnel is placed over an Erlenmeyer flask (see Figure 0.12). When the mixture is poured into the funnel, the solid collects in the filter paper while the liquid goes through into the flask. This method is most often used when the liquid is collected to be analyzed or used for further reactions.

Figure 0.12. Funnel and flask for gravity filtration.

Vacuum filtration uses a Buchner funnel, a cylindrical funnel usually made of ceramic with holes in the bottom, or a Hirsch funnel, a smaller cone-shaped funnel also often made of ceramic. The funnel is placed on a rubber adapter over a filter flask, which looks like an Erlenmeyer flask but has a port on the side from which you can pull a vacuum. When doing a vacuum filtration, a setup like that shown in Figure 0.13 is used. The second filter flask acts as a trap to ensure that no liquid from your mixture ends up in the vacuum pump, which could damage it. The stopcock vent enables you to control the strength of the vacuum and to turn it on and off. Scientists prefer this method when they aim to collect and analyze the solid, because the filtration is quick and the solid can be dried by using the vacuum to pull air through the solid.

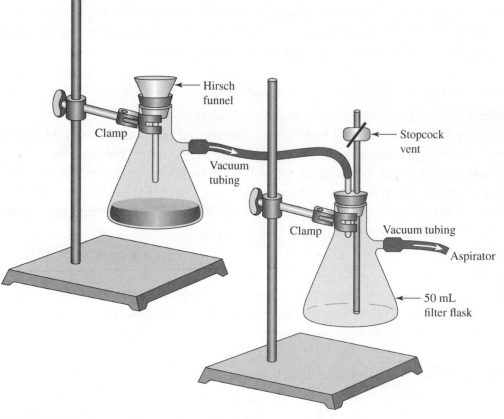

Figure 0.13. Set-up for vacuum filtration.

Recrystallization

Recrystallization is an efficient method for removing impurities from a solid. It takes advantage of the fact that most solids will dissolve to a greater extent in warm liquid than in cold liquid. The solid is dissolved in warm liquid, which is then allowed to cool off. As the solution cools, the solid crystallizes out and can be collected by filtration. If you synthesize aspirin, you may purify it by recrystallizing it. Further investigation details are given in that procedure.

Recording your data

It is important, as you perform your investigations, to take adequate notes on what you do, measure, and observe. Carefully record your procedures, write down the measurements you make (including the units!) and any observations about your investigation. Write down your predictions about the investigation, the questions you seek to answer, and the results. Record your calculations and write conclusions. Your instructor may give you a worksheet to record your results for each investigation or ask you to keep a laboratory notebook.

Data analysis

Once you have recorded your measurements you will need to look carefully at your data and observations to draw conclusions. Sometimes you can do this directly, but often you will need to further analyze your data by performing calculations or making graphs.

Calculations

Many investigations in this laboratory manual require calculations. The purpose of the calculations is usually to relate a quantity that you have measured (such as the volume of base used to titrate an acidic solution) to another quantity you are interested in but can't measure directly (the amount of acid in the solution). Details of calculations for each investigation will be given within the investigations. When doing calculations, make sure that the units of your numbers cancel out as they should, and always report the unit of your answer with the number in your lab reports.

Graphing

Often, large amounts of data are best presented as a graph. Graphs are a pictorial way of presenting information, and will allow you to see trends and relationships between the data you collect. Figure 0.14 shows how data relating mass and volume can be shown as a graph.

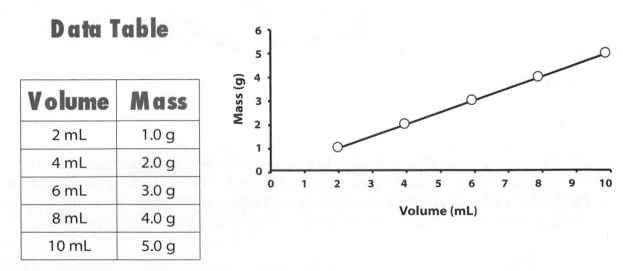

Data Table

Volume	Mass
2 mL	1.0 g
4 mL	2.0 g
6 mL	3.0 g
8 mL	4.0 g
10 mL	5.0 g

Figure 0.14. Graphical representation of data: volume as a function of mass

The graph in the figure shows that as the volume increases, the mass increases in direct proportion. Data that produce a straight line when plotted are said to have a linear relationship. Representing data with a graph makes it straightforward to estimate the value of the mass for a volume that is in-between measured data points, a process known as **interpolation**. As an example, if the volume were 5 mL, it is easy to see that the corresponding mass would be 2.5 g. It is also easy to extend the line of the graph and obtain data beyond the range of measured points, a process known as **extrapolation**. Thus, for example, if the volume were 15 mL, you should be able to convince yourself that the mass would be about 7.5 g. Care must be taken when extrapolating certain types of data, however, because the trends observed in one range of a graph are not necessarily sustained in other regions of the same graph.

A straight-line relationship such as that in Figure 13 can be summarized with a simple algebraic equation of the form $y = mx + b$. In this equation, x and y are the values for the two quantities being plotted (e.g. volume and mass), m is the **slope** of the plotted line, and b is the **y-intercept** (the value of y when the line crosses the y-axis, or the point where $x = 0$). The slope of a line can be calculated using the equation below. In this equation, $y_2 - y_1$ is the difference between the y values for two data points, and $x_2 - x_1$ is the difference between the x values for the same two data points.

$$\text{slope} = m = \frac{y_2 - y_1}{x_2 - x_1}$$

The slope summarizes the relationship between the columns of data. In the example above (Figure 14), it is easy to see that the slope is 0.5 g/mL, and it represents the relationship between mass and volume, or the density of the substance. Because the line crosses the y axis at 0, the y-intercept is 0, and the equation that represents the line is $y = 0.5x + 0$, or simply $y = 0.5x$.

While it is often useful to sketch out a graph by hand, using a computer graphing program takes much of the tedium out of the process. Many programs, for example, can create a complete graph from your data and create a linear regression line, using a statistical method to calculate the best straight-line fit for a set of data points. This can be particularly useful when there is a lot of "scatter" in the data, as in the graph in Figure 0.15.

Several investigations in this lab manual ask you to create graphs, which can be done by hand or on a computer as requested by your instructor.

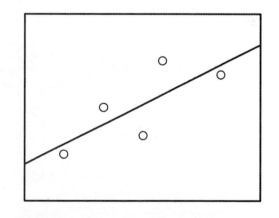

Figure 0.15. Linear regression fit of data

Notes

Preparation and Properties of Gases in Air

Asking Questions

- What is in the air you breathe?
- What gas makes up the majority of the air you breathe?
- Which gases are important to sustaining life?
- What are some of the sources of carbon dioxide in air?
- What are some differences between the air you inhale and the air you exhale?

Preparing to Investigate

In this investigation, you will prepare samples of two gases in the air, oxygen and carbon dioxide. You will then investigate some of the properties of the gases.

To prepare oxygen, you will use a catalyst to decompose a familiar household product, hydrogen peroxide (H_2O_2). In this investigation, you will use potassium iodide (KI) as a catalyst to decompose hydrogen peroxide into water and oxygen.

$$2 \ H_2O_2 \xrightarrow{\text{catalyst}} 2 \ H_2O + O_2$$

Carbon dioxide will be prepared from another common household product, baking soda, which has the chemical name sodium bicarbonate ($NaHCO_3$). When acetic acid (vinegar, $C_2H_3O_2$) is mixed with sodium bicarbonate, a chemical reaction occurs forming sodium acetate ($NaC_2H_3O_2$), which is a salt, water (H_2O), and carbon dioxide (CO_2).

$$NaHCO_3 + HC_2H_3O_2 \longrightarrow NaC_2H_3O_2 + H_2O + CO_2$$

Both gases will be generated in "zipper" plastic bags, and you will have an opportunity to make observations about the reactions. Samples of the gases, as well as samples of air and exhaled air, will be tested for flammability using a burning wood splint and for reactivity with a water solution of calcium hydroxide, $Ca(OH)_2$, also known as "limewater."

Finally, you will investigate what happens when these gases dissolve in water. In particular, you will determine whether or not they react with water to form an acid. Chapter 6 in your textbook explains concepts of acidity and pH as well as some of the effects that rising atmospheric CO_2 levels are having on the oceans. To test for changes in acidity, you will use an acid-sensitive dye, also known as an indicator, that changes color with pH. The indicator in this case, bromothymol blue, is blue in the absence of acid and yellow in the presence of acids.

Making Predictions

After reading *Gathering Evidence,* make a table listing your predictions for the tests listed in Part II. Write down what each chemical test – limewater, indicator, and wood-splint – indicates about the composition of the gas you are testing. Then, predict results for air, exhaled air, carbon dioxide and oxygen for each test.

Gathering Evidence

Overview of the Investigation

1. Prepare samples of exhaled air, carbon dioxide, room air and oxygen in plastic "zipper" bags.
2. Test these gases for reaction with limewater.
3. Test these gases for acidity in water solution.
4. Test these gases with a glowing splint.
5. Clean up.

 STOP! Safety glasses must be worn *at all times* while doing chemistry investigations.

Part I. Generating the Gases

General Instructions: The gases will be generated in plastic zippered reclosable bags, which can be easily sealed and unsealed. It is important to <u>completely</u> seal the bags. Before proceeding further, test to be sure that a sealed bag with air in it does not leak when you squeeze it gently.

A. Exhaled Air

1. Exhale through a straw that is inserted into a 1 pint heavy duty "zipper" bag. Crimp the straw, and repeat as needed until the bag is inflated with your breath.

B. Carbon Dioxide

1. Place a teaspoonful (about 2 grams) of sodium bicarbonate, $NaHCO_3$, in the bottom corner of a 1-pint heavy-duty "zipper" bag.

2. Fill two plastic transfer pipets with acetic acid (vinegar). To fill a pipet, try to squeeze nearly all the air out and then let the bulb fill with liquid. A second squeeze may help in getting more of the air out.

3. Place the pipets with the vinegar into the plastic bag with the $NaHCO_3$. Smooth out the bag so it contains a minimum amount of air; then seal the bag. (Take care not to press against the pipet.)

4. Hold the sealed plastic bag as shown in *Figure 1.1* and slowly squeeze one pipet so that the vinegar drops onto the NaHCO₃. Observe carefully what happens (several changes should be apparent) and record your observations. **Keep the bag sealed.** Once the first pipet is empty squeeze the other so that the vinegar reacts.

5. You should now have a sealed plastic bag partially filled with carbon dioxide (CO_2) gas. *Leave the bag standing upright (zipper on top) by leaning it against something on the lab bench.*

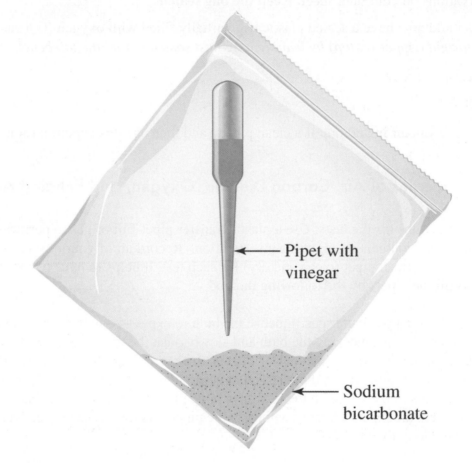

Pipet with vinegar

Sodium bicarbonate

Figure 1.1 Sodium bicarbonate in the "zipper" bag

C. Oxygen

1. Use a spatula to place a small pinch of potassium iodide (0.5 gram or less) in the bottom corner of another 1-pint heavy-duty "zipper" bag.

2. Fill two plastic transfer pipets with 10% hydrogen peroxide (H_2O_2). See directions above for filling the pipets.

CAUTION!   Hydrogen peroxide is corrosive and must be handled with care.

3. Place the pipets with 10% hydrogen peroxide into the plastic bag with the potassium iodide. Smooth out the bag so it contains a minimum amount of air and then seal the bag. (Take care not to press against the pipet.)

4. Hold the sealed plastic bag as shown in *Figure 1.1* and slowly squeeze one pipet so that the H_2O_2 drops onto the potassium iodide. Wait a few moments and then gently squeeze it also. (Try to get all of the liquid *out* of the pipets. Observe what happens and record your observations on your data sheet. **Keep the bag sealed.**

5. You should now have a sealed plastic bag partially filled with oxygen (O_2) gas. *Leave the bag upright (zipper on top) by leaning it against something on the lab bench.*

D. Room Air

1. Squeeze room air in and out of a clean pipette and then use this captured room air in Part II.

Part II. Properties of Air, Carbon Dioxide, Oxygen, and Exhaled Air

General instructions for the tests: Use a plastic transfer pipet-full of gas to perform each test on each of the gases you generated in this investigation. Record all of your observations on your data sheet. Use a fresh pipet (clean and dry) for each test. It helps to have two students working together to fill the pipets by the following method.

To fill a pipet with a particular gas, squeeze the bulb to expel as much as possible of the air inside the pipet. Keep squeezing the bulb and slowly push the tip of the pipet against the zip seal at one corner of the plastic bag containing the gas to be tested. With a bit of practice, you will be able to just push the pipet tip so that the seal opens around it. Taking care not to touch the liquid or solid chemicals in the bag with the pipet, push the pipet tip into the bag. Then release the bulb so that gas enters the pipet. Quickly withdraw the pipet. As the tip leaves the bag, immediately reseal the bag along the "zip strip."

A. Limewater Tests

1. In the first row of a wellplate, add 10 drops of limewater to the first 4 wells. Place the wellplate on a dark surface or a small piece of black paper.

2. Fill a clean pipet with carbon dioxide gas. Place the tip of the pipet into the first well. By gently squeezing the pipet, slowly bubble the gas through the limewater solution. Observe and record the results on the data sheet.

3. Repeat the test using a sample of oxygen gas bubbled into the second well. Record your observations.

4. Repeat the test using a sample of exhaled air bubbled into the third well. Record your observations.

5. Repeat the test using a sample of ordinary air bubbled into the fourth well. Record your observations.

B. Tests with an Indicator Solution

1. In the third row of the wellplate, add 10 drops of water and one drop of bromothymol blue indicator solution into each of the first 4 wells. Place the wellplate on a light-colored surface or a piece of white paper.

2. Fill a clean pipet with carbon dioxide gas. Place the tip of the pipet into the first well. By gently squeezing the pipet, slowly bubble the gas through the indicator solution. If the indicator changes color to yellow, it indicates that the solution has become acidic. Record the results on the data sheet.

3. Repeat the test using a sample of oxygen, exhaled air, and ordinary air in the next 3 wells. Record your observations.

C. Glowing Wood-Splint Test

1. Use a clean, dry pipet to obtain another sample of carbon dioxide gas.

2. With a match, ignite the end of a wood splint or toothpick (or simply use a wood match). After it has burned for a few seconds, blow out the flame. Continue to blow on the embers so that they glow.

3. Have your lab partner hold the pipet filled with carbon dioxide so that the tip is very near the glowing ember and *gently* squeeze a puff of carbon dioxide gas directly at the glowing portion. Observe and record the results.

4. Repeat the same procedure for oxygen, exhaled air, and ordinary air.

Clean-up

Clean the wellplate with soap and a brush. Discard the solutions, pipets, and the bags in appropriate waste containers as your instructor directs you. Do not put *anything* down the sink drain unless you are told that is permitted.

Analyzing Evidence

1. Describe in your own words the processes for generating oxygen and carbon dioxide gases.

2. Which of the four gases reacted with limewater?

3. What colors did you see when the four gases were mixed with indicator solution?

4. What happened to the burning wood splint when exposed to each of the four gases?

Interpreting Evidence

1. Rank the gas samples from most amount of oxygen present to least amount of oxygen present. Which test(s) from the investigation led to this ranking?

2. Rank the gas samples from most amount of carbon dioxide present to least amount of carbon dioxide present. Which test(s) from the investigation led to this ranking?

3. Does exhaled air or ordinary air contain more CO_2? How do you know?

4. Based on your indicator solution results, which gas leads to more acidic solutions?

Making Claims

What can you claim about the air you inhale and the air you exhale based on your results from this investigation?

Reflecting on the Investigation

1. According to Chapter 6 of *Chemistry in Context*, the oceans absorb 25-40% of all anthropogenic CO_2 emissions.

   a. Based on your observations in this investigation, what effect will this have on the pH of the oceans?

   b. List two consequences of ocean acidification. (Hint: see Ch. 6 of your textbook.)

2. Based on what you observed about the interaction of carbon dioxide with the glowing splint, explain how CO_2 fire extinguishers work.

3. Based on what you observed about the interaction of oxygen with the glowing splint, explain why liquid oxygen is an extremely hazardous material.

4. If your blood becomes too acidic, you may begin to hyperventilate, which causes your blood O_2 levels to increase, and your CO_2 levels to decrease. Write the equation for reaction of CO_2 with water, and explain why reducing the amount of CO_2 in the blood will reduce the acidity of the blood.

5. Are carbonated beverages acidic or basic? What do you think happens to the acidity of a carbonated soft drink as it "flattens"?

6. What is the purpose of including ordinary air in these investigations?

7. You may have indicated that nitrogen is the most abundant gas in air. Why do you think you did not perform these investigations with pure nitrogen?

Extracting Limonene with Liquid CO$_2$

Asking Questions

- What phase changes occur in CO$_2$ as temperature and pressure increase?
- Why is it important that a substance be soluble in the solvent used for extraction?
- What aspects of supercritical and liquid CO$_2$ could lead them to being classified as environmentally friendly solvents?

Preparing to Investigate

Essential oils are organic compounds extracted from natural sources such as fruits, herbs, and spices. They are used as flavorings and fragrances in a variety of products. Traditionally, essential oils are isolated through the process of steam distillation or by extraction with organic solvents. The process of **extraction** separates a compound or small number of compounds from a more complex mixture through differences in solubility. When a particular solvent is introduced, the soluble compounds dissolve in the solvent while the insoluble compounds do not. The compounds can then be separated from each other. In this investigation, you will extract a compound called *limonene* (Figure 1), a major component of the flavor and odor of oranges, from orange rind.

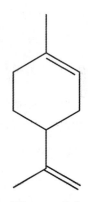

Figure 2.1. Structure of limonene

Many traditional extraction methods have serious environmental drawbacks, including intensive energy use and the production of large amounts of toxic waste. The past few decades have led to a large increase into the investigation and use of supercritical CO$_2$ in place of organic solvents. This solvent is nonflammable, relatively nontoxic, and environmentally benign. Additionally, the properties of the solvent can be altered and used for a variety of applications by simply changing the temperature and pressure of the CO$_2$.

We are most familiar with CO$_2$ in the form of a gas, as it exists in the atmosphere or as we exhale it. Some of you may have seen solid CO$_2$, which is also known as dry ice. Unlike water ice, which melts to a liquid, dry ice **sublimes** which means it goes directly from a solid to a gas. The reason why can be seen in the phase diagrams below.

Phase diagrams describe the form a substance takes at different temperatures and pressures. Figure 1B shows the phase diagram for water. If you draw a line at atmospheric pressure (1 atm) from -100°C to 400°C, you'll notice that solid water becomes a liquid (at 0°C) and then a gas (at 100°C) as the temperature is increased. If at 0°C you were to reduce the pressure, you reach the **triple point**, where water exists in all three phases. At pressures below the triple point, liquid water cannot exist, and ice would sublime directly to water vapor. If you were to heat the water to 374°C and raise the pressure above 218 atm, you would reach the **critical point**. Above this temperature and pressure you have a **supercritical fluid**, where the sample exhibits both gas and liquid characteristics. It will completely fill a container like a gas but will dissolve substances as a liquid does. Supercritical fluids occur naturally at underwater ocean vents and in the atmospheres of gas giant planets such as Jupiter and Saturn.

In the phase diagram for carbon dioxide (Figure 1A), you can see that the triple point occurs at 5.1 atm and -57°C. This means that atmospheric pressure is less than the triple point pressure, so liquid carbon dioxide cannot exist at atmospheric pressure. This is why dry ice sublimes. You will also notice that the critical point for CO_2 occurs at a much lower temperature and pressure than that of water – 31°C and 7.4 atm, which makes it quite easy to generate. Therefore, supercritical CO_2 has replaced more toxic solvents in applications such as dry cleaning and decaffeination of coffee.

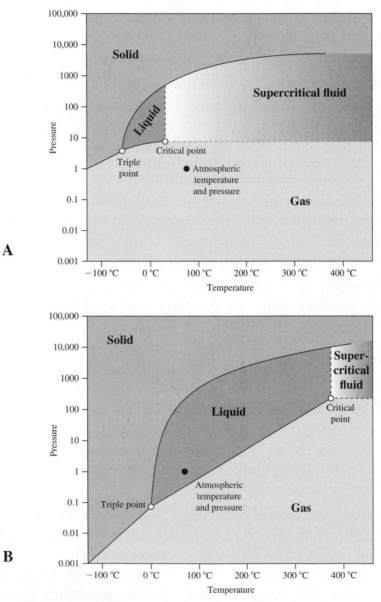

Figure 2.2. Phase diagrams for (A) carbon dioxide and (B) water.

In order to form liquid CO_2 as part of this investigation, we must raise the pressure of the carbon dioxide above the triple point pressure. We will do so by sealing dry ice inside a plastic tube, where the pressure from the subliming CO_2 will rise high enough to generate a liquid. You will use the liquid CO_2 to extract limonene from orange rind. CO_2 is a nonpolar solvent, so it is appropriate to use it to dissolve non-polar compounds such as limonene. In fact, what you will isolate is the essential oil of orange, a mixture of compounds, of which greater than 90% is limonene.

Making Predictions

- Predict what percentage of the initial mass of rind will be recovered as oil.

- After reading *Gathering Evidence*, prepare a data sheet that includes space for your predicted percent recovery, the initial mass of the tube, the mass of the added orange peel, the mass of the tube with extracted limonene, the mass of the extracted limonene, and measured percent recovery. Leave room for your observations and calculations of the final mass and percent recovery.

Gathering Evidence

Overview of the Investigation

1. Prepare your centrifuge tube by making and inserting a coiled copper wire.
2. Grate orange peel and place it on top of the coil.
3. Fill the tube with dry ice, cap it, place in cylinder of warm water, and watch the extraction.
4. Record your observations and measure the amount of limonene produced.
5. Clean up.

Part I. Preparing the Sample

 STOP! Safety glasses must be worn *at all times* while in the chemistry laboratory. Do NOT use any glass containers during this investigation.

1. Using a balance, measure the mass of a clean, dry 15-mL plastic centrifuge tube. Record the mass on your data sheet.

2. Grate the colored part of the peel of half an orange using the smallest grating surface of a cheese grater. Weigh the peel using a balance, and record the mass on your data sheet. It should be approximately 2.5 g.

3. Use a 20-cm piece of copper wire to make a trap for the solids by coiling the wire tightly at one end so that it fits in the tube at the top of the taper (see Figure 3). This will prevent your solid from falling to the bottom of the centrifuge tube and interfering with recovery

of the essential oil. The other end of the wire should extend upward but not past the top of the tube. The wire should be completely inside the tube so that the cap can be sealed.

4. Place your orange rind sample into the tube, making sure that none of it falls to the bottom of the tube. Do not pack it tightly.

Part II. Extracting the oil

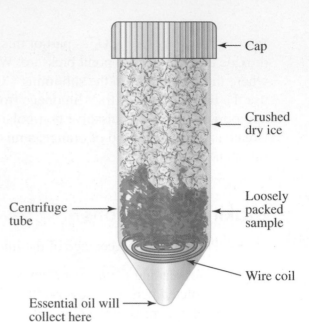

Cap

Crushed dry ice

Loosely packed sample

Centrifuge tube

Wire coil

Essential oil will collect here

NOTE: Due to the safety issues involved with the rapid increase of pressure during this procedure, it is important to read and understand the entire extraction procedure (steps 1-8) before beginning!

Figure 2.3. Investigation setup.

1. Fill a clear plastic cylinder about two-thirds full of warm tap water. Place the cylinder on the bench top and move any items that should not get wet away from the cylinder. Splashing may occur if the top of your reaction tube shoots off the tube during the extraction.

    CAUTION! Do not heat the water in the cylinder or add hot water later in the procedure. Any sudden increase in temperature of the surrounding water when the tube is under pressure can cause the cap to blow off suddenly and violently.

2. Use a thermometer to measure the temperature of the water, and record it on your data sheet. Remove the thermometer from the cylinder before continuing.

3. Fill the centrifuge tube to the top with crushed dry ice. Tap the tube on the bench and continue to add more dry ice until the tube is full.

    CAUTION! Wear gloves when handling dry ice because direct contact can damage skin tissues.

4. Twist the cap on tightly until it stops turning. **Note:** If the cap never stops turning or otherwise appears not to fit the tube well, do not proceed. You will need to use a new cap or a new centrifuge tube.

5. Immediately after capping, drop the centrifuge tube, tapered end down (cap up), into the water in the cylinder. Pressure will begin to build in the tube.

CAUTION!   Watch the extraction from the side, not the top! If the tube shatters or the cap shoots off any projectiles will be directed straight up. The plastic cylinder functions as a secondary container and protects you from possible injury.

6.   After 15 seconds, liquid CO$_2$ should be visible. The entire extraction will take approximately 3 minutes. If no liquid has appeared after 1 minute, there is not a sufficient seal on the tube. Carefully remove the tube from the cylinder, tighten the cap, and put the tube back in the cylinder. If repeated trials don't produce liquid, try replacing the cap and/or the tube. *Never remove the tube from the plastic cylinder when the CO$_2$ is liquid.*

7.   After the liquid has evaporated and the gas is no longer escaping, remove the tube from the cylinder and open the cap.

CAUTION!   Open the tube slowly and only after the gas has escaped. Do not point the cap at anyone, including yourself, as you are opening it.

8.   Repeat the extraction once or twice more by adding dry ice to the tube as explained in step 3 above and repeating the subsequent steps. After two or three extractions you should see a small amount of liquid at the bottom of the tube. This is the essential oil.

9.   After your final extraction, carefully remove the coiled wire and the solid. If any solid remains in the tube, remove it with a spatula. Dry the outside of the tube with a paper towel. Measure and record the mass of the tube and the oil on your data sheet, along with any observations about the oil. Your instructor will demonstrate the proper way to observe the odor of the oil. Also measure and record the temperature of the water in the cylinder.

Clean-up

Discard or recycle the contents of your test tubes as directed by your instructor. Clean your thermometer with soap and water to remove all traces of the oil.

Analyzing Evidence

1.   Calculate the mass of the extracted oil by subtracting the mass of the empty centrifuge tube from the mass of the tube with the oil.

Mass of essential oil = (mass of tube with oil) – (mass of empty tube)

2.   Use the calculated mass of oil to determine how much oil is present in a particular mass of rind (percent recovery). Calculate the percent recovery of your essential oil by dividing the mass of the oil by the mass of the orange rind that you put in the tube.

$$Percent\ recovery = \frac{mass\ of\ essential\ oil}{mass\ of\ orange\ rind\ used} \times 100\%$$

Interpreting Evidence

Answer these questions in paragraph form and include in your written report.

1. Explain the differences between the *liquid* CO_2 that you generated in this investigation and *supercritical* CO_2 that is used as a commercial solvent.

2. Why was it necessary to add warm water to the cylinder that acted as a secondary reaction vessel? Consider that dry ice sublimes at -78°C at 1 atm. What is the minimum temperature at which liquid CO_2 can be generated?

3. How did the temperature of the water change during the extraction? Did it increase or decrease? Explain your observations.

4. What was your percent recovery? Does this seem low or high for an extraction of essential oil? Why?

5. Identify several factors that could have affected your yield of oil and explain whether they would have made the yield higher or lower.

Making Claims

What can you claim about the use of carbon dioxide to extract essential oil from a natural product?

Reflecting on the Investigation

Answer these questions in paragraph form and include in your written report.

1. Limonene is a member of a class of molecules called *terpenes*, which are composed of one or more five-carbon *isoprene* units. Terpenes are important natural products, used in cleaning products and as organic pesticides. Some vitamins and other important biomolecules are also terpenes.

   a. How many isoprene units are present in limonene? (Hint: count the carbons)

   b. Another common and commercially important terpene is menthol (shown at right), a major component of the essential oil of peppermint. You may be familiar with this compound as a major ingredient of Vicks Vapo-Rub and similar products. How many isoprene units are present in menthol?

   c. Retinal is a light-sensitive compound that plays an important role in vision. The formula for this compound is $C_{20}H_{28}O$. How many isoprene units are found in this compound?

   d. Our bodies make retinal from beta-carotene, the orange compound found in carrots and other vegetables. The chemical formula for beta-carotene is $C_{40}H_{56}$. How many isoprene units does it contain?

e. If you have covered Ch. 10 in *Chemistry in Context*, identify the functional groups present in limonene and menthol.

2. The introduction to this investigation states that carbon dioxide is a benign solvent, but you have probably read in *Chemistry in Context* that CO_2 is a greenhouse gas that can lead to global warming and contribute to ocean acidification. Normally, no new CO_2 is produced in order to make dry ice and supercritical CO_2. Rather, it is produced as a byproduct of other processes. Search the web to find out where it comes from, and explain why it does not generally contribute additional CO_2 to the atmosphere.

3. Supercritical CO_2 has replaced chlorinated organic solvents such as methylene chloride and perchloroethylene (PERC) for applications such as dry cleaning and decaffeination of coffee. These solvents were initially chosen for their ability to dissolve a wide variety of substances and because they are not flammable, making them relatively safe to work with compared to hydrocarbon solvents. Search the web for these chlorinated solvents, identify some of their hazards, and explain why CO_2 may be a better option. You may wish to search for the Material Safety Data Sheet (MSDS) which records important information about chemical safety.

4. The use of liquid and supercritical carbon dioxide is a large area of research for green chemists. Read the introduction to green chemistry in this book and take a look at the key ideas in green chemistry inside the front cover of your textbook. Identify which key ideas apply to the use of supercritical CO_2 as an industrial solvent.

Reference:

Investigation adapted from McKenzie, L. C.; Thompson, J. E.; Sullivan, R.; Hutchison, J. E. *Green Chemistry*, **2004**, *6*, 355-358.

Notes

Chromatographic Study of Dyes and Inks

Asking Questions

- How can you determine whether a substance is pure or is a mixture?
- What properties can be used to separate components in a mixture?
- Do similar dyes and inks contain the same components?
- What aspect of the inks in "washable" markers would enable them to be separated under the conditions of this investigation?

Preparing to Investigate

How can we know if something is pure, containing just one element or compound, or a mixture of several substances? Chapter 1 in *Chemistry in Context* introduces the concepts of pure substances and mixtures, and notes that most of the materials we encounter daily, such as air, water, and food, are mixtures of chemical substances.

In this brief exploration, you will investigate artificial dyes used in drink mixes and/or inks from washable felt-tip pens to find out whether the colors are produced by a pure substance or a mixture of substances. By careful comparison, it should be possible to determine whether some of the substances have colored components in common.

This investigation uses **chromatography**, a common method for separating and studying the components of a mixture. Chromatography is described in detail in the *Laboratory Methods* section, which you should read before doing this investigation. In chromatography, the components of a mixture are allowed to move along a stationary surface. Each component in a mixture retains its own properties and moves at a rate determined by its own characteristics. Since they move at different rates, the components become spread out and separated from each other like runners in a foot race. When paper is the stationary surface, the usual strategy is to place small spots of substances near one edge of the paper. That edge is placed in a liquid, such as water, and the liquid moves up the paper by capillary action. The liquid carries the components with it, but at differing rates depending on their solubility in water.

Making Predictions

- Suggest some reasons why some components of the dyes will move up the filter paper at different rates.
- After reading *Gathering Evidence*, prepare a data sheet to record your observations. Be sure to leave room for the paper chromatography strips and to identify the solvents.

Gathering Evidence

Overview of the Investigation

1. Place spots of colored dyes or inks on a piece of filter paper.
2. Immerse one edge of filter paper in water and allow water to move the spots upward.
3. Compare the spot locations for the various inks.

Part I. Separating the Components of Dyes and Inks

These investigations should be done individually so that each person has a completed chromatogram, but you are encouraged to compare results and discuss the interpretation. Your instructor will tell you which procedure to follow, or you may do both.

A. *Testing Food Dyes*

1. Obtain four drink mixes (e.g., Kool-aid) of different colors.

2. If solutions of the drink mixes have not been prepared, follow this procedure to make a concentrated solution of each. Use a balance to weigh 0.5 g of each powder and place each mix into a separate test tube. Your instructor will explain the correct use of the balance, if necessary. Add 6 drops of water to each test tube and shake the test tube. Add water one drop at a time just until the powder is dissolved.

 STOP! Do not drink any beverages that have been opened or used in the laboratory.

3. Obtain a rectangular piece of filter paper, approximately 5 cm x 12 cm. Use a *pencil* to write your name at the top and to draw a line 2 cm from the bottom edge of the paper, as shown in *Figure 3.1*. Make four small pencil marks 1 cm apart along the line.

4. Using a small capillary tube, place a small spot of one of your solutions on the first mark. Allow the spot to dry and repeat the spotting.

5. Make spots with the other solutions on the other marks, using a new capillary tube for each solution. In pencil, label the name of the solution spotted on the mark.

6. Lay a glass rod across the top of a 400 mL beaker. Fold or roll the top edge of the filter paper so that it hangs on the rod inside the beaker with the bottom of the filter paper close to but not touching the bottom of the beaker, as shown in *Figure 3.1*.

7. Carefully add a small amount of water so that there is a shallow pool of pure water in the beaker. The bottom of the paper should extend into the water, but the water level should not be higher than the line that contains the spots. Leave the beaker undisturbed on the desk and observe what happens.

8. When the top edge of the water has moved about 2/3 of the way up the paper, remove the paper and lay it on a paper towel to dry. While the paper is still damp, draw a line *in pencil* to show the top edge of the wet area.

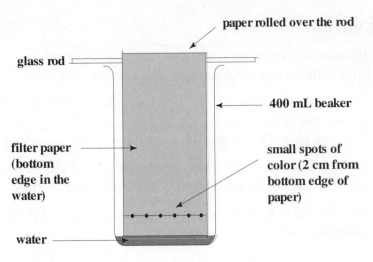

glass rod

paper rolled over the rod

400 mL beaker

filter paper
(bottom
edge in the
water)

small spots of
color (2 cm from
bottom edge of
paper)

water

Figure 3.1 Set-up for chromatography investigation.

B. Water-soluble Inks

1. Obtain four pens that contain different colors of vivid, washable inks. Good choices are the Vis-à-Vis overhead-projector pens, LiquidMark Washable Markers, or any other markers labeled "washable".

2. Follow steps 3 through 8 above. However, rather than using a capillary tube to make your spots, you can use the pens directly to draw a small spot on the filter paper.

3. Replace the water with another solvent, such as methyl alcohol, ethyl alcohol, or rubbing alcohol (which is a mixture of 70% isopropyl alcohol and 30% water). Do the investigation the same way, and record your observations.

Part II. Optional Variations or Extensions

1. Investigate what happens with other kinds of pens. Some possibilities: other transparency pens, "Sharpie" or "Flair" pens, and ballpoint pens. (For ballpoint pens, make sure you have a very dark spot.) Use a fresh sheet of filter paper for each investigation. For permanent markers or ballpoint pens, try replacing the water with a solution of alcohol or acetone in water.

2. Using a circular piece of filter paper, cut a 1-cm wide strip to the center of the paper disk. Spot six colors around the rest of the filter paper about 1 cm from the center of the circle. Fold down the paper tab that you cut earlier down into the center of a 50-mL beaker that contains a enough water that it comes in contact with the tab. The circle of the filter paper should rest easily on the rim of the beaker. As the water moves up the tab and out through the filter paper circle, the colors will fan out towards the edge of the filter paper in a floral pattern.

3. Try doing the investigation with a natural substance, such as pigments from plants or flowers. You can extract the pigments from the leaves or petals of the plants by grinding in a mortar with rubbing alcohol, and then filtering the resulting solution. Spot the filter paper using a capillary tube as described above.

Analyzing Evidence

1. When the filter papers are completely dry, staple or tape each one to your data sheet. Include a brief summary of what samples and solvent were used.

2. Using the formula given in the *Laboratory Methods* discussion of thin-layer chromatography, calculate the R_f for each spot on your filter paper. How do the R_f values compare? Can you identify any spots that have the same R_f values?

3. Look at your data and determine which of the dyes or inks appear to contain a single colored substance and which are mixtures. Describe your evidence.

4. Look at your data and determine which samples have colored components in common with other samples. Explain your reasoning.

Interpreting Evidence

1. From your results, predict which colored components are most soluble in water and which are least soluble in water. Explain your reasoning with specific examples from your data.

2. Which of your samples had the highest number of components? Explain any trends in the data that you see.

Making Claims

What can you claim about the similarities and differences in the dyes and inks that you sampled?

Reflecting on the Investigation

1. Describe an investigation you could do in your kitchen to convince a friend that the hard coating on an M&M contains a mixture of several food colorings. Explain what you would do and what you would see as results.

2. Why was it important to use a pencil to label the filter paper?

3. Suggest a possible explanation for why the results using alcohol are different from those in water.

4. (If you did optional step 1) An interesting household problem is that of removing ballpoint pen stains from clothing. In light of your observations, can you suggest what will and will not likely work? What additional tests could you do to investigate this problem further?

Graphing the Mass of Air and the Temperature of Water

Asking Questions

- Air has a mass because it is composed of chemical compounds. What are the compounds that you'll be weighing today?

- What is the relationship between the mass and pressure of air in a closed system?

- Do you expect the graphs you'll construct to have linear or curved regressions? Why?

- What are some advantages of using graphs to represent data?

Preparing to Investigate

In this assignment, you will collect some investigation data about air and water that lends itself to presentation in graphical form. You will then learn how to construct graphs in ways to make the visual presentations most effective.

Two short laboratory exercises are included. Your instructor will specify whether you are to do one or both. The first exercise studies the relationship between the pressure and mass of air in a closed container. The second investigates the relationship between time and temperature as a hot liquid cools.

Graphs present information pictorially, allowing you to see trends and relationships between the data you collect. The *Laboratory Methods* section describes in detail how to prepare graphs and analyze your data using **interpolation** and **extrapolation**. For this investigation you will do a regression analysis of your data. Carefully read the section on graphing in the *Laboratory Methods* section before doing the investigation.

Making Predictions

- After reading *Gathering Evidence*, sketch a graph for each exercise that you predict will reflect the relationship between the data sets. Identify whether the expected regression is linear or curved and explain why you predict that relationship.

- Prepare a data sheet that starts with the predicted graphs and includes space for observations and data tables for the assigned exercises.

Gathering Evidence

Overview of the investigation

Part I. Weighing Air

1. Obtain an empty 2-liter bottle with a tire valve mounted in the cap.

2. Inflate the bottle to about 40 pounds per square inch with a bicycle pump.

3. Weigh the bottle and measure the pressure in the bottle.

4. Release some of the air; measure the pressure and weigh the bottle again.

5. Repeat Step 4 several times until all of the air pressure has been released.

Part II. Cooling Water

1. Heat some water in a beaker to a temperature of about 80°C.

2. Allow the water to cool and record its temperature at regular time intervals.

 STOP! Safety glasses must be worn *at all times* while doing chemistry investigations.

Part I. Weighing Air

1. For this investigation, you will use a laboratory balance for measuring the mass of an object. Your instructor will explain the use of the particular balances in your laboratory. If you are using an electronic balance, it is important to make sure the balance reads zero when empty (usually accomplished by pressing a TARE button). Check the zero each time you use the balance. This investigation will require you to make measurements to the nearest 1/100 of a gram (0.01 gram).

2. Obtain a clean, dry, plastic soda bottle (2-liter size is preferred) with a screw cap that has been fitted with a tire valve. Put a thermometer strip into the bottle and tightly screw on the cap.

3. Record the temperature of the air inside the bottle. Each time you manipulate the bottle, record the temperature from the inner thermometer. Also record your observations including whether the bottle feels warm or cool to the touch.

4. Become familiar with using the tire gauge to measure the air pressure. Your instructor can assist you, if necessary. Make sure you can push the gauge on squarely without letting out much air and can read the pressure units on the protruding stem. The pressure scale is probably marked in pounds-per-square-inch (psi). Atmospheric pressure is 14.7 psi, and the tire gauge measures pressures over (not under) atmospheric pressure.

5. Attach the hose from a hand tire pump to the valve. Place the bottle in a protective enclosure (such as a wastebasket). Pump air into the bottle until the pressure on the tire gauge reads about 40 psi.

6. Measure and record the mass of the bottle. Wait 5 minutes and measure the mass again.

7. If the mass of the bottle (plus air) is the same as it was initially, the cap is sealed tightly, and you can start to collect data. If the mass of the bottle decreased by more than 0.05 g, tighten the cap and repeat steps 4 and 5.

8. If necessary, pump air into the bottle until the pressure reads 30–40 psi. Carefully measure the air pressure in the bottle using the tire gauge and immediately record this in the data table. Weigh the bottle on a balance and record the mass (to the nearest 0.01 gram) and temperature.

9. Let a small amount of air out of the bottle. Measure and record the pressure and temperature again and then measure and record the mass.

10. Repeat step 8 until you have at least five sets of pressure, temperature and mass data for points between your highest pressure and 10 psi.

11. Let all of the pressurized air out of the bottle, and measure and record the mass and the temperature. Remember that the bottle is not empty but now contains air with a pressure of around 14.7 psi.

12. **OPTIONAL:** These measurements can be done very rapidly. If time permits and you are interested, you might like to try repeating the study. Your data may be more accurate because measurement technique often improves with repeated investigations.

Part II. Cooling Water

 STOP! This part of the investigation involves an open flame. No flammable chemicals should be in the vicinity. Long hair should be tied back, and loose sleeves on clothing should be avoided.

1. Use a thermometer or temperature probe to measure the air temperature in the room. Record this on your data sheet.

2. Measure 50 mL of water using a graduated cylinder, pour it into a 100 mL beaker, and place the beaker on a ring stand over a Bunsen burner.

3. Light the Bunsen burner, heat the water to about 80°C, and then turn off the Bunsen burner.

4. Allow the water to cool to 75°C without stirring or removing the beaker from the ring stand.

5. When the water has cooled to 75°C, start recording the temperature at intervals of 2 minutes. Use a wall clock, wrist watch, or stop watch to keep track of the time and record the temperatures to the nearest 0.5 degree.

6. Continue making and recording measurements until the water has cooled to about 35°C.

Analyzing Evidence

A. Graphing Conventions

1. Graphs should be neat, legible, and well-organized.

2. The title should be descriptive and tell the reader what has been plotted, such as "Volume of plastic as a function of its mass." Titles such as "M vs. V" do not give enough information and should not be used.

3. The horizontal axis (x axis) and the vertical axis (y axis) should be clearly labeled to show what is plotted (e.g., volume) and the units (e.g., mL).

4. The *scale* of the graph should be chosen so that the graph fills as much of the paper as practical. In general, the scales on the x axis and the y axis do not need to start at 0. *Figures 14* and *15* in the *Laboratory Methods* section show examples of correctly scaled graphs. The data points fill the graph space quite effectively. *Figure 3.1* below shows three examples of incorrectly scaled graphs.

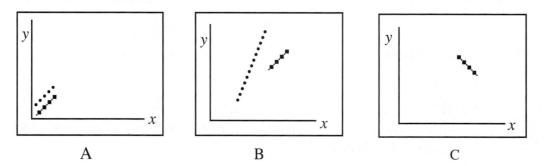

A   B   C

Figure 4.1 Examples of graphs with incorrect scales

Graph A: The increments on the x and y axes should be made smaller to spread the graph out over the page.

Graph B: The starting point on the x axis should be changed and the increments on the x axis should be made smaller so the graph takes up most of the page.

Graph C: The starting point on both the x and y axes should be changed and the increments on both axes should be made smaller so the graph takes up most of the page.

5. If the data points appear to represent a linear (straight-line) relationship, use a ruler to draw a single straight line that best represents the average relationship. If the data points seem to follow a curve rather than a straight line, then draw the best *smooth* curve

through them. It is unlikely that all of the points will fit exactly on one smooth line (either straight or curved); therefore, some judgment must be exercised in deciding on the "best" fit. You are trying to identify a *trend* in the data, so it is not effective to draw zig-zag lines that connect the data points.

6. When some sets of investigation data are plotted, it becomes apparent that certain points seem "out of line" with all of the others. Scientists often call these "outliers." In this case, it is likely that an error was made, either in the original measurement or in writing it down. (It is also possible that an error was made in placing the point on the graph—this is the first thing to check.) If a point really seems to be out of line with the others, it is appropriate to exclude it when deciding on the best line. Clearly, this requires some judgment.

B. Graphing the Data

Part I. Weighing Air

1. Identify the range of recorded masses from lowest to highest. Select a *convenient* set of values that will include this range of masses. Use these to label the vertical (*y*) axis on the graph. You may want to select values with subdivisions of 10 since the masses are in decimal units.

2. Similarly, identify the range of recorded pressures. These data should be from 0 to about 30 or 40 psi. Choose a convenient set of values for the horizontal (*x*) axis and label the axis.

3. Carefully plot the investigation points using a pencil. Make small dots, and then draw small circles around them so that they show up clearly.

4. Examine the data points carefully. If the data points appear to be in a straight line, use a ruler or other straight edge (preferably transparent) to draw the "best" line through the data points. If the data points seem to follow a curve rather than a straight line, then draw the best *smooth* curve through them. Don't do this too hurriedly—it requires some judgment. If the points are somewhat scattered, there should be equal numbers of them on either side of the line.

5. If specified by your instructor and if the line is straight, calculate the slope of the line. To do this accurately, DO **NOT** use particular data points. Instead, select two places on the line near the opposite ends of the line. Carefully read off the *x* and *y* values for each, and then use the equation given in the *Preparing to Investigate* section to calculate the slope.

Part II. Cooling Water

1. Identify the range of recorded temperatures from highest to lowest. Select a *convenient* set of values that will include this range of temperatures. Use these to label the vertical (*y*) axis on the graph.

2. Select a convenient set of values for the horizontal (*x*) axis, starting at zero, to cover the number of minutes over which you made measurements. Label this axis.

3. Carefully plot the investigation points using a pencil. Make small dots, and then draw small circles around them so that they show up clearly.

4. Examine the data points carefully. If the data points appear to be in a straight line, use a ruler or other straight edge (preferably transparent) to draw the "best" line through the data points. If the data points seem to follow a curve rather than a straight line, then draw the best *smooth* curve you can through them. Don't do this too hurriedly—it requires some judgment. If the points are somewhat scattered, there should be equal numbers of them on either side of the line.

Interpreting Evidence

1. An interesting further use of the data set from *Weighing Air* is to extrapolate the line to -14.7 psi. (If a computer-graphing program is available, this becomes easy to do simply by changing the scale on the horizontal axis.) What do you think you will see?

2. Suppose you could pump some air out of the bottle rather than pumping it in. Predict how the mass would change. Is there a limit to the change?

3. All of your mass measurements were actually the combined mass of the bottle itself plus the air it contained. How could you find out the mass of the air alone? (Hint: What is the significance of the weight of the bottle at -14.7 psi?)

4. Suggest a possible explanation as to why the bottle warmed and cooled when it did.

Making Claims

What can you claim about the relationship between the mass and pressure of air in a closed system? What can you claim about the relationship between temperature and time for water as it cools? Use the data and the graphical representations to support your claims.

Reflecting on the Investigation

1. Write a paragraph describing the advantages of using a graph to represent data rather than just using a data table. Include examples encountered in this investigation's introduction and results.

2. Figures 3.5 and 3.6 in your textbook show two plots of well-known data describing the increase in atmospheric carbon dioxide with respect to time. Study these figures and answer the following questions.

    a. What two quantities are plotted against each other in each graph? What are the units?

    b. The data from Mauna Loa is included in the plot spanning thousands of years. Why is it useful to have this data plotted separately, as in the inset graph in Figure 3.6?

    c. People continue to argue about the validity of extrapolating this data to predict future atmospheric carbon dioxide levels. In what ways might an extrapolation be valid, and what concerns may be raised by trying to extrapolate this data?

3. Choose another graph from your textbook, write down the figure number, and answer the following questions.

    a. What quantities are plotted against each other in the graph? What are the units?

    b. Describe what this graph tells you about the relationship between these quantities (e.g., does one quantity increase or decrease as the other increases?). Does the graph indicate correlation, causation, or no relationship between these quantities? (Look up *correlation* and *causation* if you don't know what these terms mean.)

Notes

What Protects Us From Ultraviolet Light?

Asking Questions

- What materials do the best job of protecting us from UV light?
- What properties of a material lead it to reduce our exposure to UV light?
- What information would be important to know when designing a material to protect us from UV light?

Preparing to Investigate

In this investigation, you will design and conduct an investigation to test the ability of materials to protect us from ultraviolet (UV) light. We have become increasingly aware of the need to protect ourselves from exposure to UV light. Fortunately, this awareness has led to the increased development of materials that can reduce our exposure to the dangerous components of sunlight while allowing us to enjoy time outside.

As described in Chapter 2 of *Chemistry in Context*, exposure to UV light can be dangerous. In particular, UV light has sufficient energy to cause changes in DNA that can lead to skin cancers. In recent years, depletion of the ozone layer has allowed more UV light to reach the surface of the earth, increasing the risk of skin cancer. One strategy to protect ourselves from UV light would be to avoid exposure to any type of light. Few among us would be willing to take this to the extreme of moving underground, but we've all moved under a shady tree or used a hat to block out light.

A wide variety of commercial lotions are available with sun protection factors (SPF) ranging from 2 to over 50. Many of these sunscreens contain organic molecules, such as oxybenzone and octyl methoxycinnamate, that absorb light specifically in the ultraviolet region of the spectrum. Some sunscreens contain very small particles of titanium dioxide or zinc oxide that are designed to scatter the light. The lenses in sunglasses often absorb UV-A (320–400 nm), UV-B (280–320 nm), and even UV-C (100-280 nm) light.

To measure UV exposure, you will use beads that change color when interacting with UV light. The more UV light they are exposed to, the faster their color will change. By measuring how long it takes the beads to change color, you can determine the amount of UV light reaching them. You will be provided with a variety of protective materials to test. Your instructor will give you a list of what is available and may allow you to bring additional items from home.

You will collect your data outdoors because you need to use unfiltered UV light. The beads are quite sensitive, though, so direct sunlight is not required and working in a shady spot usually gives better results. As you conduct your investigation, remember to protect the beads from all sides so that you are only measuring the exposure through the material. In between trials, put the beads in a dark place, such as your pocket. They will quickly change back to their original color when not exposed to UV light.

Remember that good investigation design requires including control variables and multiple trials. As you collect data, change only one variable at a time. You will also want to determine how the beads respond when left totally unprotected. Repeating measurements will allow you to report average data instead of single data points.

Making Predictions

Rank your proposed materials in order of their predicted ability to protect the beads from UV light and explain why you chose this order.

Gathering Evidence

Overview of the Investigation

1. Design an investigation that uses at least three protective materials and includes appropriate controls.

2. Get your instructor's approval and suggestions for your procedure.

3. Conduct your investigation.

4. Report your findings.

Part I. Designing the Investigation

1. With your partner or other team members, develop a plan for your investigation. Outline the procedure that you intend to follow on your own paper.

2. Check to see that you have included

    a. at least three protective materials,

    b. descriptions of controls, and

    c. a plan for multiple trials that states what, how and when you will collect data.

3. Obtain instructor approval to proceed. (Your instructor may offer some suggestions or ask questions to help you refine your investigation plan.)

Part II. Conducting the Investigation

1. Follow your investigation plan.
2. Keep a careful written record of what you do and note any changes from your original plan.
3. Record your data in an appropriate table.

Part III. Reporting the Findings

Prepare a written report that includes the following sections. Write each section in paragraph form, unless directed otherwise, and be sure to include the answers to the questions in that section.

1. Asking Questions

    a. What information were you trying to learn from your investigation? Be specific about and include your choices of materials and methods.

    b. Why is your investigation important?

2. Making Predictions

    a. Report about your ranking of your proposed materials in order of their predicted ability to protect the beads from UV light and explain why you chose this order.

3. Gathering Evidence

    a. What method did you use to gather evidence? (A numbered list may be the best format for reporting this information.)

    b. Did you have any changes to the original plan? If so, why?

4. Analyzing Evidence

    a. Include your data tables with measured and average values

    b. Identify which of the trials contained investigation controls and explain why they were important to include.

5. Interpreting Evidence

    a. Answer the questions listed in *Interpreting Evidence* below.

6. Making Claims

    a. Use the questions in *Making Claims* below to generate claims that you can make from the evidence you collected.

7. Reflecting on the Investigation

    a. Explain any investigation difficulties and what you did to solve them.

    b. Describe one specific change that you would make in your procedure if you were doing the investigation again. Explain how this change could lead to better or more interesting results.

    c. Answer the questions listed in *Reflecting on the Investigation* below.

Analyzing Evidence

Complete your data tables and calculate average data. Include the data in your written report.

Interpreting Evidence

Include the answers to these questions in paragraph form in your written report.

1. What materials do the best job of protecting us from UV light? Rank the materials from least protective to most protective and explain what evidence you used to determine the order.

2. Were your predictions correct about each material's ability to protect from UV light? Explain any differences between your predictions and your results.

3. What properties of a material cause it to protect from UV light? Which of these properties might be most useful in the development of a UV protector for skin? Which could be used in developing a UV protector for large or immovable objects?

4. What is the purpose of an investigation control? Describe the controls you used and what information they provided for you.

5. Why are multiple trials of the same procedure needed? Give an example of an undesirable result that could have occurred if you had not used multiple trials.

Making Claims

Use these questions to generate claims. Include them in paragraph form in your written report.

- What can you claim about the materials you tested and their ability to protect from UV light?
- What general conclusions does your evidence suggest about interactions between materials and UV light?

Reflecting on the Investigation

Include the answers to these questions in paragraph form (except for #6) in your written report.

1. Explain any investigation difficulties and what you did to solve them.

2. Describe one specific change that you would make in your procedure if you were doing the investigation again. Explain how this change could lead to better or more interesting results.

3. Compare the energy of ultraviolet light photons to that of infrared and visible light photons. Explain why UV light is potentially more dangerous than visible or infrared light.

4. What information would be important to know when designing a material to protect us from UV light?

5. Most sunscreen lotions and sunglasses claim to protect against UV-A and UV-B light. Why don't they mention UV-C light? Is it less dangerous than the other types of UV light? What happens to the UV-C? (See Chapter 2 of *Chemistry in Context*.)

6. UV-visible absorption spectra of several materials are shown in *Figure 5.1* on the next page. Each spectrum shows the wavelength range(s) at which the material absorbs light. If a material has a large absorbance at a particular wavelength, it will protect us from light of that wavelength. If absorbance is low, the material will not provide effective protection.

    a. <u>On each spectrum</u>, label these types-of-light regions along the *x*-axis:

    > UV-A light 320–400 nm
    >
    > UV-B light 280–320 nm
    >
    > UV-C light 100–280 nm

    b. Examine the spectrum of glass. Which types of light (UV-A, UV-B, and/or UV-C) does glass absorb best? From what type of UV light does this glass fail to protect us? Explain.

    c. Now look at the spectrum of clear plastic wrap. What types of light does the plastic wrap absorb? Would plastic wrap be good for preventing sunburn or skin cancer? Explain.

    d. What types of light does the sunscreen absorb?

    e. Finally, look at the spectrum of cotton fabric. How is this different from all the other spectra? Does the fabric selectively absorb certain wavelengths? What is the fabric doing that is different?

Figure 5.1   Ultraviolet Absorption Spectra of Protective Materials

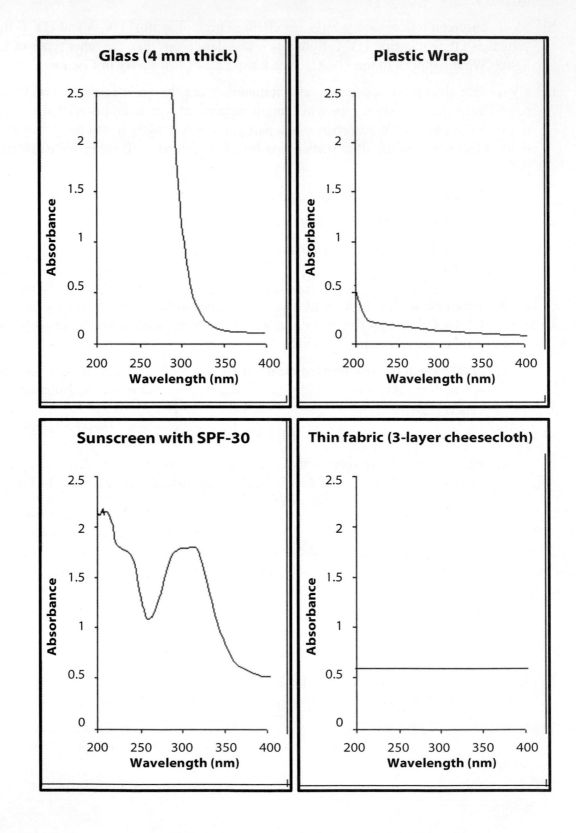

Color and Light

Asking Questions

- Why are some objects colored? How are they interacting with light?
- What is spectroscopy?
- What is the relationship between the color of an object and the color(s) of the light that it absorbs?

Preparing to Investigate

We have all seen white light separated into its colored components in rainbows and light passing through a prism. These colors of light can be detected by human eyes and therefore are part of what is called the visible region of the **electromagnetic spectrum** (see Chapter 2 in *Chemistry in Context*). Each color has a unique position in the electromagnetic spectrum that is identified by its **wavelength**. In the visible region, the wavelengths are in the range of billionths of a meter and are expressed in **nanometers**, nm (1 nm = 1 x 10^{-9} meter).

In this investigation, you will investigate the separation of visible light and interactions of light with colored solutions. After determining the colors of different wavelengths of light in the visible region, you will measure the wavelengths of visible light that are transmitted or absorbed by colored solutions. You will construct a **spectrum** for each substance that shows how much light of different wavelengths is absorbed or transmitted.

You will use a **spectrophotometer** for this investigation. The components and use of a spectrophotometer are described in detail in the *Laboratory Methods* section of this lab manual, and you should read that section before proceeding to the investigation. The operation of each spectrophotometer is different, and so you will be given detailed directions from your laboratory instructor for the operation of your specific instrument. Spectrophotometers are sensitive and expensive instruments and so it is important to follow the instructions carefully so that the equipment is not damaged.

Making Predictions

- Which of the solutions, the red, blue, or green, do you expect to absorb light at the shortest wavelength? Longest wavelength? Explain why you predict that relationship.

- After reading *Gathering Evidence*, prepare a data sheet that includes data tables for the spectroscopic measurements and space for observations, predictions and graphical representations.

Gathering Evidence

Overview of the Investigation

1. Prepare a piece of chalk to reflect light from the spectrophotometer source.
2. Determine the color of light of selected wavelengths from 400 to 700 nm.
3. Measure the transmittance (or absorbance) spectrum for a red solution and a blue solution.
4. Predict and then measure the spectrum for a green solution.
5. Plot a graph of transmittance (or absorbance) vs. wavelength for each colored solution.
6. Optional: Use this procedure to investigate a colored substance of your own choosing.

Part I. Assigning Wavelength to Color

1. Turn on the spectrophotometer and let it warm up. Your instructor will let you know whether to choose the %T or A mode, and show you how to adjust the wavelength.

2. Take a half-inch long piece of chalk and rub it on the blackboard until one end is worn down to a forty-five-degree angle (*Figure 6.1*). Place the chalk in the test tube and insert it into the spectrophotometer so that the slanted side is pointing toward the light source.

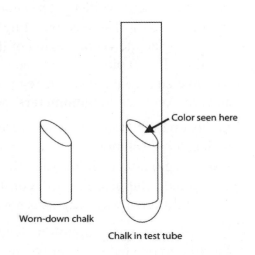

3. Adjust the wavelength to 500 nm.

4. Look down the tube to the slanted piece of chalk. You should see a colored band of light. If not, raise, lower, or rotate the test tube slightly until you do. If you still cannot see the colored band, consult your instructor.

5. Slowly adjust the wavelength in both directions and observe that the color of the light band changes.

Figure 6.1 Chalk prepared for the spectrophotometer

6. If you change the wavelength far enough in either direction, you will not see any color because the wavelength will be beyond the region that your eye can detect. By careful observation, find out the limits for *your* eyes and record these limits in your observations.

7. Finally, find out which wavelengths correspond to which colors. To do this, return the wavelength to 400 nm. Look down the tube and record in your data table the color you see. Change the wavelength to 440 nm and again record the color you see. Repeat this at 40 nm intervals from 400 to 720 nm.

Part II. Collecting Transmission or Absorbance Data for Red and Blue Solutions

1. Obtain three matched test tubes or other containers, often called cuvettes, that fit your spectrophotometer. Fill one of them about half full with a red dye solution. Fill another test tube or cuvette with a blue dye solution. Fill the last tube half full with pure water. The water will be the blank.

2. Put the blank in the instrument and, if necessary, adjust the zero on the spectrophotometer.

3. Remove the blank, insert the sample tube with the red solution and close the cover. *Without changing any knobs or pressing any buttons*, carefully observe the meter reading and record this in the data table as % T (or A) for the red solution at 400 nm.

4. Place the third tube containing the blue solution in the instrument, close the cover, and read the % T (or absorbance) and record in your data table.

5. Change the wavelength dial to 420 nm. Repeat steps 3 and 4 by first inserting the blank, closing the cover and adjusting the meter to read 100% T or 0 A. Then insert the samples, observe the meter reading, and record the % T or absorbance. Repeat with the blue solution.

6. Proceed in the same fashion, at 20-nm intervals, to 700 nm. Although it may sound tedious, data collection goes quite rapidly when two students work together. Remember that *each* time you change the wavelength, you must insert the distilled water blank and adjust to 100% T or 0 A. Then collect the data for both the red and blue solutions.

8. Finally, make a *prediction* of what you think the spectrum will look like for a green solution, using the data from the red and blue solutions. Show it to your instructor before proceeding. Then, rinse and fill the sample tube with a green solution and proceed to record the transmittance or absorbance data for this solution in the same manner as for the other solutions. Make sure to use the blank for each measurement.

Part III. Spectrum of an Indicator Solution

1. Rinse out the sample test tube and fill it halfway with the blue indicator solution provided (bromthymol blue). A small amount of sodium hydroxide has been added to this solution in advance to be sure the solution is alkaline. Rinse and half-fill another test tube or cuvette with the same indicator solution. Add a drop of hydrochloric acid solution to make the solution acidic. It should change color to yellow.

2. Proceed in the same manner as in Part II. At each wavelength chosen, first insert the blank test tube and set to 100% T or 0 A. Then measure the colored solutions from 400 nm and to 700 nm. Be sure to measure both test tubes (acidic and basic) before proceeding to the next wavelength.

3. Find a way to make an intermediate form of the indicator solution that is green. You will have to investigation by adding small amounts of very dilute acid (hydrochloric acid) and very dilute base (sodium hydroxide) to achieve a green color. Predict what the spectrum will look like, then measure the spectrum of this green solution and compare it with the spectra of the blue and yellow forms.

Part IV. Investigating Other Colored Solutions

There are many colored substances that you could use in a spectrophotometric study. Some common examples include beverages, such as Kool Aid, soft drinks, or juices, water solutions of jello, water-color paints, felt-tip pen inks, food color dyes, and hair dyes. You could even study small pieces of colored glass if they can fit into the sample chamber. Whatever you choose must be transparent (meaning that you can see through it) and fairly bright, and distinctly colored.

You may have to investigation to find a suitable dilution of your substance so that the lowest transmittance does not drop below 5% T (or absorbance above 1.5). Then use the skills you have learned previously to collect data and plot a spectrum of the substance.

Part V. Collecting Ultraviolet or Infrared Spectra

Your instructor may demonstrate, or allow you to use, a spectrophotometer that measures ultraviolet or infrared spectra. You may be able to obtain an ultraviolet spectrum of a sunscreen material of the type used in Investigation 5. You also may be able to obtain an infrared spectrum of a greenhouse gas such as carbon dioxide.

Analyzing Evidence

1. Plot the data for each colored solution or substance. It is convenient to do this by hand on graph paper but you can also use a computer program. The wavelength should be on the horizontal axis (*x* axis), ranging from 400 nm at the left edge to 700 nm at the right edge. Percent transmittance or absorbance should be on the vertical axis (*y* axis); % T should range from 0 at the bottom to 100 at the top or absorbance should range from 0 at the bottom to 1.0 (or 2.0) at the top.

2. First plot the red solution. *Use a pencil so that corrections can be made.* Carefully plot the data, making a small point with a circle around it for each measurement. When you are finished, draw a *smooth* curved line through the points. The line may not touch every point.

3. Repeat this procedure with other colored substances. You can plot more than one spectrum on the same graph, but, if more than one spectrum is on as single graph, label each spectrum clearly or use a different color pencil for each line.

Interpreting Evidence

1. How do the spectra change when measuring red, green, and blue solutions?

2. At what wavelengths does each solution have its greatest transmittance (lowest absorbance)? What color(s) of light correspond to these wavelengths?

3. At what wavelengths does each solution have its lowest transmittance (highest absorbance)? In these regions, the dye molecules are absorbing the light. What color(s) of light is the red/blue/green dye solution absorbing?

Making Claims

What can you claim about the color of an object and the color(s) of the light that it absorbs?

Reflecting on the Investigation

1. Humans can typically see light in the range of 400-700 nm. However, some animals can see at wavelengths outside of this range. For instance, bees and some spiders can see ultraviolet light, while snakes and other reptiles have vision extending to the infrared wavelengths. What adaptive advantages might this extended vision offer these animals?

2. Why is the sky blue? And why does it appear red or orange at sunrise and sunset?

3. Our eyes see color by using light-sensitive cones. We have three types of cones that each see a different color – red, green and blue. If we have only these three types of cones, how can we see more than just three colors? People who are color-blind lack one or more type of cones. Explain how vision is affected for someone who lacks red cones.

4. Mixing colored light is different than mixing colored paint. For instance, if you mix red and yellow paint, you get orange, and if you add blue to the orange you get brown or black. However, your computer screen and television make colors by mixing light. Mixing red and green light makes yellow light, and adding blue light makes white. Investigate and write a paragraph about how and why mixing light is different than mixing paint.

Notes

Testing Refrigerant Gases

Asking Questions

- How do refrigerant gases work?
- Which compounds were traditionally used as refrigerant gases before chlorofluorocarbons (CFCs)?
- What properties of CFCs make them ideal refrigerant gases?
- Why were CFCs outlawed in the 1990s?
- Which compounds have taken the place of CFCs as modern refrigerant gases?

Preparing to Investigate

Refrigerators and air conditioners take advantage of the physical properties of liquids and gases. Heat energy is given off when a vapor condenses to a liquid. Conversely, heat energy is absorbed when a liquid vaporizes. The cooling mechanism in a refrigerator or air conditioner is a closed system containing a compound that boils at a low temperature but that can be converted easily back to a liquid under pressure. The liquid is allowed to vaporize in metal tubes inside the refrigerator. As it vaporizes, it takes heat from its surroundings and cools the inside of the refrigerator. The vapor produced is then pumped to metal tubes on the rear of the refrigerator, outside the cold compartment, and it is converted back to a liquid using pressure applied by a compressor. As the gas is condensed, it loses the heat it picked up inside the refrigerator and makes the coils on the back of the refrigerator feel warm. Through this process, heat is pumped out of the refrigerator and into the kitchen. In an air conditioner, the heat removed from the air in a room is pumped to the outside.

To efficiently carry out this refrigeration cycle, refrigerant gases must have certain physical properties. They must have boiling points well below 0°C so that they do not condense back to a liquid inside the refrigerator or freezer compartment. They must be easily liquefied under pressure. Also, they should be non-flammable. Before CFCs, the most common refrigerant gases that met these requirements were ammonia (NH_3), which boils at –33°C, and sulfur dioxide (SO_2), which boils at –10°C. Both can be liquefied at room temperature under moderate pressure. At room temperature (75°F), ammonia will turn to a liquid at a pressure about 10 times greater than atmospheric pressure, and sulfur dioxide will liquefy at about 5 times atmospheric pressure.

Today, it is hard to understand why the discovery of CFCs was hailed as one of the wonders of modern chemistry. These compounds are now viewed with great concern because of their destructive effect on the stratospheric ozone layer (see Ch. 2 in *Chemistry in Context*). In this investigation, you will utilize chemical reactions to prepare samples of ammonia gas and sulfur dioxide gas and will compare their properties to those of a modern refrigerant gas. Since CFCs are no longer available, you will use one of the HCFC or HFC gases that have replaced CFCs.

Sulfur dioxide will be prepared by reacting sodium sulfite with hydrochloric acid, as in the following reaction equation:

$$Na_2SO_3 \ + \ 2 \ HCl \rightarrow 2 \ NaCl \ + \ H_2O \ + \ SO_2$$

Warming a concentrated solution of ammonium hydroxide will produce ammonia:

$$NH_4OH \ (aq) \ + \ heat \rightarrow \ H_2O \ + \ NH_3$$

Making Predictions

After reading *Gathering Evidence*, make a table to record your data. How do you think the properties you will measure will compare between the gases?

Gathering Evidence

 CAUTION! Do not do this investigation if you have a sensitivity to sulfites or have asthma. Sulfites have been shown to cause severe asthma attacks in a few sensitive individuals. See your instructor before the lab for an alternate assignment.

Overview of the Investigation

1. Construct a reaction vessel for generating sulfur dioxide.
2. Prepare a sample of sulfur dioxide and observe its properties.
3. Generate ammonia gas and describe its properties.
4. Collect a sample of refrigerant gas and describe its properties.

Part I. Constructing a Reaction Vessel for Sulfur Dioxide

Note that five different sizes of plastic pipets are needed for this investigation. They are shown in Figure 7.1.

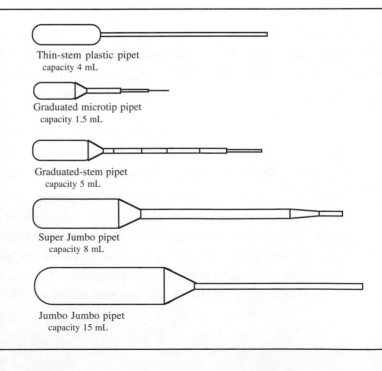

Thin-stem plastic pipet
capacity 4 mL

Graduated microtip pipet
capacity 1.5 mL

Graduated-stem pipet
capacity 5 mL

Super Jumbo pipet
capacity 8 mL

Jumbo Jumbo pipet
capacity 15 mL

Figure 7.1. The types of transfer pipets used in this investigation.

1. Cut off the bulb section of an 8-mL super-jumbo pipet just after the stem widens to the size of the bulb (Figure 7.2). Discard the stem end, and stand the cut-off bulb in a well of a wellplate.

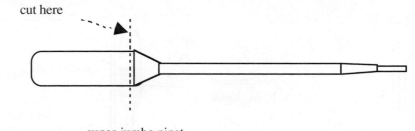

cut here

super-jumbo pipet

Figure 7.2. Where to cut the super-jumbo pipet.

2. Cut off the stem off of a 15-mL jumbo-jumbo pipet (Figure 7.3). Discard the bulb end.

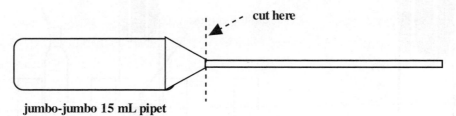

cut here

jumbo-jumbo 15 mL pipet

Figure 7.3. Where to cut the jumbo-jumbo pipet.

3. Obtain a #0 two-hole rubber stopper. Push the stem of the 15-mL jumbo-jumbo pipet through one hole, and a graduated micro-tip pipet through the other hole (Figure 7.4).

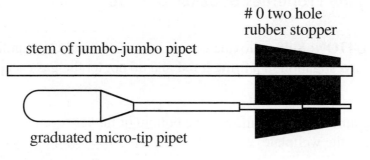

0 two hole
rubber stopper

stem of jumbo-jumbo pipet

graduated micro-tip pipet

Figure 7.4. The top of the reaction vessel.

4. The stopper assembly should fit tightly into the bulb section of the super-jumbo pipet prepared in step 1 (Figure 7.5). The reaction vessel will just fit into a well in a 24-well wellplate (Figure 7.6).

5. Cut off the bulb of a graduated-stem pipet at the 1-mL mark. This will serve as the container to collect the sulfur dioxide. Place this bulb in a well next to the reaction vessel. The entire assembly should look like Figure 7.6.

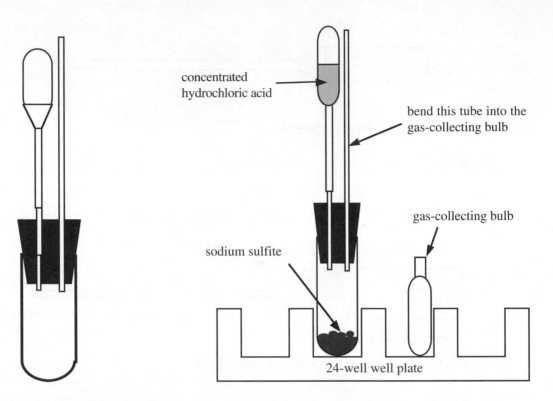

Figure 7.5. The reaction vessel.

Figure 7.6. The complete reaction set-up for generating sulfur dioxide.

Part II. Testing the Properties of Sulfur Dioxide

 CAUTION! Sulfur dioxide is very toxic and should be handled cautiously. Perform this investigation in a **fume hood** and do not let the gas escape into the lab.

1. Place 1 gram of sodium sulfite in the bottom of the reaction vessel, and place the vessel in a well of the wellplate.

2. Remove the small pipet from the stopper and fill it with concentrated hydrochloric acid. Then carefully put it back in the stopper without squeezing out any of the acid.

 CAUTION! Hydrochloric acid is corrosive and must be handled with care. Wear gloves when handling it and notify your instructor immediately of any spills.

3. To generate and collect sulfur dioxide, bend the straight pipet stem so that the end goes into the gas-collecting pipet bulb. Gently squeeze the bulb of the pipet containing hydrochloric acid to add 2-3 drops to the sodium sulfite. Sulfur dioxide gas will be produced and will carry over into the bulb.

4. To test water solubility: Add about 40 mL of water to a 50-mL beaker. Generate and collect some sulfur dioxide, then remove the straight stem from the gas-collecting bulb and quickly invert the bulb into the beaker of water so that the opened end is below the surface of the water. Observe for a minute or two. If water comes up into the bulb, the sulfur dioxide is soluble in water. After recording your observations, rinse out the gas-collecting bulb with water.

5. To test water reactivity: Put a few drops of water and a drop of bromothymol blue indicator into the gas-collecting pipet bulb. Bend the straight pipet stem so that the end goes into the gas-collecting bulb. Generate some sulfur dioxide and then shake the gas-collecting bulb gently and record any changes in the color of the indicator. Bromothymol blue is yellow in acid and blue in base. Rinse your gas-collecting bulb with water after you have recorded your observations.

6. To test reactivity with an oxidant: Add a few drops of pink potassium permanganate ($KMnO_4$) solution to the gas-collecting bulb, re-insert the straight pipet stem and collect some more sulfur dioxide. Shake the collecting bulb and record your observations. Note that potassium permanganate is colored but that its reaction products are usually colorless.

7. To observe the odor: If in the course of doing the investigations you have smelled the sulfur dioxide, describe the odor of the gas in your observations. *Do not put your nose near the pipet stem and take a deep breath! Sulfur dioxide is toxic and you should not inhale too much of it!*

8. To clean up: Put 50 mL of water in a 250-mL beaker. Remove the pipet with concentrated hydrochloric acid from the reaction vessel and slowly add the acid to the water. Then, add solid sodium bicarbonate a little at a time until it no longer bubbles. Dispose of the neutralized acid, the remaining sodium sulfite, and the potassium permanganate solution as directed by your instructor. Your gas generation apparatus should be rinsed and then saved or discarded as your instructor directs you.

Part III. Testing the Properties of Ammonia

 CAUTION! Ammonia is very toxic and should be handled cautiously. Perform this investigation in a **fume hood** and be careful to not let the gas escape into the lab.

1. In the fume hood, use a hot plate to heat 150 mL of water in a 250-mL beaker to 65°C.

2. Fill a thin-stem plastic pipet half full of concentrated ammonium hydroxide and turn it upside-down. The stem of the pipet must not contain any liquid. If liquid is in the stem, force it to the top by gently squeezing the pipet bulb and blot off the liquid with a paper towel.

3. Cut off the bulb of a graduated-stem pipet to use as a gas-collection bulb. Place it over the stem of the transfer pipet containing ammonium hydroxide (Figure 7.7).

4. Dip the end of the bottom pipet into the hot water. After a few seconds the ammonium hydroxide should begin to bubble as ammonia gas is driven off the warm ammonium hydroxide solution. After 25-30 bubbles appear, your gas-collection bulb should be full of ammonia gas. Remove the pipet from the hot water as you test your gas samples.

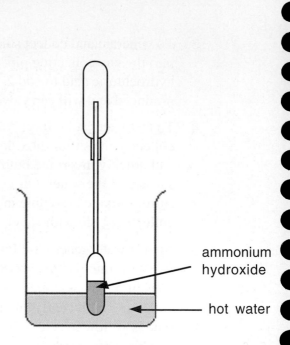

ammonium hydroxide

hot water

Figure 7.7. The set-up for generating and collecting ammonia gas.

 CAUTION! The solution should not be heated for more than 15 seconds at a time or the bulb will burst!

5. <u>To test water solubility and reactivity:</u> Before collecting ammonia gas, put 20 mL of water and five drops of phenolphthalein solution in a 50-mL beaker before you collect some ammonia gas. After collecting the gas, quickly remove the collection bulb, keeping the open end pointed down. Touch the opened end of the collection bulb to the phenolphthalein solution so that the open end of the bulb is about an eighth of an inch below the surface of the liquid. Gently move the bulb from side to side while the opened end is in the liquid. Describe what you observe. Note that phenolphthalein is colorless in acid and red in base.

6. <u>To test reactivity with an oxidant:</u> Before collecting your ammonia gas, put 20 mL of water and 10 drops of pink potassium permanganate solution in a 50-mL beaker. Collect the ammonia gas as described above, quickly remove the collection bulb, and test the gas with the potassium permanganate in the same manner as you did with the phenolphthalein. Record your observations.

7. <u>To observe the odor:</u> If in the course of doing the investigations you have smelled the ammonia, describe the odor of the gas in your observations. *Do not put your nose near the pipet stem and take a deep breath! Ammonia is toxic and you should not inhale too much of it!*

8. <u>To clean up:</u> Dispose of all solutions in appropriate waste containers, as directed by your laboratory instructor.

Part IV. Testing the Properties of a Modern Refrigerant Gas

These tests will be done with one of the compounds currently available as replacements for CFCs. Your instructor will provide you with the name and formula of the compound and will demonstrate how to obtain small amounts of gas from a pressurized container.

1. Cut off the stem of a graduated-stem pipet to make another gas-collection bulb.

2. Fill the dry gas-collecting bulb with refrigerant gas, keeping the opened end of the bulb pointed up.

3. Perform the same tests on the modern refrigerant gas as you did on sulfur dioxide. Test for water solubility by inverting the bulb into a beaker of water, for water reactivity by observing reaction with bromothymol blue indicator, and for reactivity with an oxidant by reacting with potassium permanganate. If possible, observe the odor of the gas. You will need to obtain a fresh sample of gas for each test.

4. To clean up: Dispose of all solutions in appropriate waste containers as instructed.

Analyzing Evidence

1. Which of the three gases is most soluble in water?

2. Which of the gases are acids, bases, or neutral compounds?

3. What do the test results from reaction with potassium permanganate tell you about the chemical properties of the gases?

Interpreting Evidence

1. When you were collecting the gases, why were you instructed to hold the collection bulb with its open end down to collect ammonia, but the open end up to collect sulfur dioxide and the modern refrigerant gas?

2. Do some online research into the toxicity of the modern refrigerant gas you used. How does its toxicity compare to sulfur dioxide and ammonia?

Making Claims

What can you claim about the various refrigerant gases that have been used?

Reflecting on the Investigation

1. On the basis of your observations, why do you think that CFCs were considered a major advancement to civilization when they were discovered?

2. Some large commercial refrigeration units still use ammonia as a refrigerant gas. Why do you think ammonia is still being used?

3. If ammonia and sulfur dioxide were used in household refrigerators, they would have to have warning labels. Write a proposed warning label for each of these gases.

Molecular Models, Bonds, and Shapes

Asking Questions

- What are some of the ways that the structure of molecules can be shown?
- How does the number of electrons shared between atoms lead to differences in bonding?
- What is the octet rule and why is it important in determining molecular shape?
- Why do some molecules not follow the octet rule?

Preparing to Investigate

In this exercise, you will have the opportunity to apply your understanding of electron arrangements in molecules and the resulting shapes of the molecules by constructing simple ball-and-stick models for some common molecules. You will investigate a number of small molecules containing carbon, nitrogen, oxygen, and hydrogen, as well as a few molecules containing fluorine, chlorine, or sulfur. The properties of chemical compounds are directly related to the ways in which atoms are bonded together into molecules. In the process of doing this exercise, you will see how models help chemists understand and predict chemical properties.

The existence of chemical compounds with fixed composition implies that the atoms in compounds must be connected in characteristic patterns. Although early models showed the atoms hooked together like links on a chain, modern representations are abstract and often mathematical in nature. It is possible, though, to represent molecular structures with reasonable accuracy by using relatively simple models. The models serve as a three-dimensional representation of an abstract idea. Molecular model building has proven so useful that it is rare to find a chemist who does not have a model kit close at hand.

An important part of this exercise involves identifying the *three-dimensional shapes* of molecules. Molecules have certain shapes depending on their component atoms and the ways in which they are bonded to each other. The chemical bonds that hold atoms together in molecules generally consist of *pairs of electrons* shared between two atoms. Electrons not involved in bonding are termed *unshared electrons*. Chapter 3 of *Chemistry in Context* shows how the three-dimensional shapes of molecules are related to the bonding. The connection between electrons and bonds are often shown through Lewis dot structures (see Section 2.3 of *Chemistry in Context*).

The important shapes encountered in this exercise are *linear, bent, triangular, pyramidal,* or *tetrahedral.* Several factors contribute to determining molecular shape:

- Electron pairs try to keep as far away from each other as possible, while still remaining "attached" to atoms.

- Electron pairs tend to be symmetrically arranged around each atom in a three-dimensional manner.

- Electron pairs *not* involved in the bonding ("unshared pairs" or "lone pairs") are equally as important as shared (bonding) electron pairs (shared pairs) in determining the overall molecular shape and arrangement of atoms.

Atoms tend to share outer electrons in such a way that each atom in the union (except hydrogen) has a share in an *octet of electrons* in its outermost shell. This generalization has come to be known as the **octet rule**. (For further information, review the discussion of Lewis structures and the octet rule in Section 2.3 of the text.) The location of each element in the periodic table provides information about the number of electrons in the outermost level of the atoms. Carbon, for example, is in Group 4A and has four outer electrons; thus, it must share four additional electrons from other atoms in order to achieve eight outer electrons (an octet). Oxygen, in Group 6A, has six outer electrons and shares two electrons from other atoms in order to achieve an octet. Hydrogen is a special case, needing to share its one electron with only one electron from another atom in order to achieve the stable outer electron configuration of the nonreactive element helium (He). This is summarized in Table 8.1.

Table 8.1 Electron Configurations in Atoms and Molecules

Atom	**Outer electrons**	**Electrons shared with another atom (bonds formed)**
Carbon	4	4
Nitrogen	5	3
Oxygen	6	2
Fluorine & Chlorine	7	1
Hydrogen	1	1

A *single bond* consists of one shared pair of electrons, a *double bond* shares two pairs (i.e., 4 electrons), and a *triple bond* is three shared pairs (6 electrons). On paper, the bonds are represented by single, double, or triple lines, respectively (–, =, ≡). In most model kits, straight sticks represent single bonds, and pairs or triplets of curved sticks or springs are used for double and triple bonds.

Making Predictions

- After reading *Gathering Evidence*, predict the shape of each molecule that you will build with the model kit. Identify why you predict each shape.

- Prepare a data sheet that includes columns for the molecule name, predicted shape, reason for prediction, total number of outer electrons, the Lewis structure, and geometry of the atoms.

Gathering Evidence

Overview of the Investigation

1. Identify atoms and number of shared electrons in a variety of molecules.
2. Use a molecular model kit to build the molecules.
3. Look at the models and describe the bonding and shape of the molecules

Building the Molecular Models

A. Getting Acquainted with the Model Set

Each pair of students should have a model set to use to build models for each of the molecules listed on the data sheet.

1. Identify which color balls represent which atoms.

2. Note the holes in the various colored balls and their positions.

3. If there are sticks of several lengths or shapes, determine which are for single and which are for multiple bonds. If you have single bond sticks in two lengths, the short ones are for bonds involving hydrogen and the longer ones are for other single bonds.

B. Using the Model Set to Build Molecular Models

Follow this procedure to build a model for each of the following molecules:

Methane – CH_4, Water – H_2O, Ammonia – NH_3, Oxygen – O_2, Nitrogen – N_2,

Carbon dioxide – CO_2, Ozone – O_3, CFC-12 – CF_2Cl_2 . CFC-22 – CHF_2Cl,

Sulfur dioxide – SO_2, Carbon monoxide – CO, Formaldehyde – H_2CO.

For more challenge, try these two molecules: Nitric oxide – NO, Nitrogen dioxide – NO_2.

1. Set aside the atoms required to build the molecule. For example, you need one carbon and four hydrogen atoms to make the CH_4 molecule.

2. Identify the number of outer electrons in each atom of the molecule by finding the element group in the periodic table that contains each atom. The number of the group is

equal to the number of outer, or valence, electrons for that atom. Write these numbers in your data table.

3. Determine how many pairs of electrons are in the molecule by dividing the *total* number of outer electrons by 2. Record this number in your data table and set aside a stick to represent each pair of electrons. For example, H_2O has one outer electron from each hydrogen atom and six outer electrons from the oxygen atom. The total number of pairs of electrons is 4 (8 divided by 2). Therefore, you will need four sticks to represent all the electrons in H_2O, two sticks for the bonds between H and O and two for the unshared electron pairs on oxygen.

4. With the collected parts, assemble the model in such a way that each atom except hydrogen has a share in an octet of electrons. If you do not appear to have enough sticks (electron pairs) to give each atom (except hydrogen) an octet, try sharing more electrons by forming double or triple bonds (replace straight sticks with curved sticks or springs). **NOTE:** If there is only one atom of one element in the molecule and more than one atom of another element, the single atom usually goes in the center of the molecule. This is the case in CO_2, but there are a few exceptions to this rule (such as N_2O, which has the arrangement NNO).

Analyzing Evidence

1. View each model. Use it and the data of number of electrons and pairs to determine the Lewis structure. Draw this in your data table for each molecule.

2. Draw the geometry of the atoms for each molecule. Determine the molecular shape (linear, bent, triangular, pyramidal, or tetrahedral) and include this information in your data table.

Interpreting Evidence

Include the answers to these questions in paragraph form in your written report.

1. Compare your predicted shapes to the geometries determined by the model. Are they the same? If not, what rules did you apply through the modeling exercise that led to different shapes from your predictions?

2. The tetrahedral shape is one of the most fundamental shapes in chemical compounds. How would you describe it in words to someone who has never seen it?

3. The octet rule is a very important rule governing the structures of molecules. Based on your work with molecular models, provide a simple explanation for the importance of eight electrons.

4. Explain in your own words how non-bonded electron pairs help determine the shapes of molecules.

5. CO_2 and SO_2 have very similar formulas. Did you find that they have the same geometry? Explain why or why not with consideration of the numbers of total outer electrons. What did you find about the geometry of SO_2 and O_3? Explain similarities or differences based on their outer electrons.

6. Predict the shapes of (a) NF_3, (b) H_2S, and (c) Cl_2O. Which molecular models from this investigation helped you to make each prediction?

Making Claims

Use these questions to generate claims. Include them in paragraph form in your written report.

- What can you claim about the molecules that you investigated and the dependence of their shape on the arrangement of outer electrons?

- Do all of the assigned molecules obey the octet rule? If not, why (or in what way) did the octet rule fail? What can you claim about the position of the atoms on the periodic table that do not obey the octet rule?

Reflecting on the Investigation

Include the answers to these questions in paragraph form in your written report.

1. Models do not necessarily have to be physical objects. They can be two-dimensional drawings or even mental constructs. Cite one or more examples of such models encountered outside of chemistry. Can you think of models that are used in your own field of study or that you will use in your future career?

2. We often use computer-generated images to help us understand molecular structures. Consult your instructor to find out what molecular viewing programs are available at your institution. View some of the molecules that you studied in this investigation, and rotate them using the computer program. What advantages do you see for viewing molecules this way as compared to the pictures in your textbook? What advantages and disadvantages do you see for these computer images as compared to the physical models you constructed in lab?

3. You will revisit many of these molecules as you go through the course. Answer the following questions about some of the molecules you studied in this investigation.

   a. Define the term "greenhouse gas" and explain how CO_2 functions as one. What other molecules that you studied today can act as greenhouse gases?

   b. What role does O_3 play in the atmosphere? What molecules are involved in its formation?

   c. Explain the problems with the use of CFCs as refrigerants. What class of molecules has replaced them?

   d. What physiological role does NO play?

Notes

Measuring Molecular and Molar Mass

Asking Questions

- How can we measure the mass of a molecule?
- Do all gases have the same molar mass?
- What are some the relationships between mass and volume of gases?

Preparing to Investigate

Chapter 3 of *Chemistry in Context* introduces **molar mass**, the mass (in grams) that contains 6.02×10^{23} (Avogadro's number) atoms or molecules of an element or compound. This investigation provides an opportunity to measure the molar masses of several gases and compare your investigation results with the accepted values for those gases.

Equal volumes of gases at the same temperature and pressure contain equal numbers of molecules. This means that comparing the masses of equal volumes of two gases is the same as comparing the masses of equal numbers of molecules of the substances. The *ratio* of the masses of an equal number of molecules therefore will be the same as the ratio of the masses of the *individual molecules* and the ratio of the *molar masses* of the substances involved. This can be written as

$$\frac{\text{mass of } n \text{ molecules of gas A}}{\text{mass of } n \text{ molecules of gas B}} = \frac{\text{mass of 1 molecule of gas A}}{\text{mass of 1 molecule of gas B}} = \frac{\text{mass of 1 mole of gas A}}{\text{mass of 1 mole of gas B}}$$

Weighing a gas is more complicated than weighing a small solid object for many reasons. First, the masses of small volumes of gas are not large and, therefore, even very small errors can impact the results. Also, objects float if the air that is displaced by the object weighs more than the object displacing it. This phenomenon, buoyancy, is easily seen in a helium-filled balloon. The balloon floats in air, and, if a helium balloon were placed on a balance, it would appear to weigh nothing! Objects that do not float still are affected by an upward force equal to the mass of air they displace and proportional to the volume of air that is displaced. For example, although your mass, the amount of matter you contain, is the same in air or in a vacuum, you would appear to weigh about 100 grams (about 1/4 lb) less in air then you would weigh in a vacuum, depending on your volume. In order to get an accurate weight for an object, the mass of the air displaced by the object must be added to the weight of the object in air. For solid objects, the correction is so small that it can be ignored in most cases. However, when weighing gases, this buoyancy effect is significant.

In this investigation, you will weigh equal volumes of different gases. You then will correct the data for buoyancy to determine the mass of each gas sample. Finally, you will use the relationships between mass and volume of gases to relate the corrected masses to a standard reference substance (oxygen, O_2). You will then be able to determine values for the molar masses of the gases that you have investigated.

Making Predictions

- Without doing any calculations, rank the following gases from lowest to highest molar mass: oxygen, methane, carbon dioxide, argon, and nitrogen. Explain why you put them in this order.

- After reading *Gathering Evidence*, prepare a data sheet that starts with the predicted ranking and includes space for observations (including temperature and pressure in the laboratory) and measurements. Also, make a table for collection of data for each of the listed gases and an unknown. Some possible data for the table might include weight of gas assembly, measured mass of gas, mass of air displaced, corrected mass of gas, calculated (or investigation) molar mass, and accepted molar mass.

- Using values for atomic weights from the periodic table, calculate the molar mass of each gas you will measure in the investigation to the nearest whole number. Enter these values into the table on your data sheet as accepted molar masses for the gases.

Gathering Evidence

Overview of the Investigation

1. Prepare a plastic bag to use as a gas container.
2. Fill the container with several different gas samples, including one unknown gas.
3. Weigh the container filled with each of the gases.
4. Determine the volume of the container.
5. Determine the mass of a liter of air at room temperature and pressure.
6. Calculate the measured and corrected mass of each gas.
7. Calculate the molecular mass and volume for each gas.
8. From the data, propose a possible formula for the unknown gas.

Part I. Weighing the Gases

Notes about weighing: The most critical measurements in this Investigation are weights (of the gas container plus gases) that are obtained by means of a **laboratory balance**. The use of a balance is explained in detail in the *Laboratory Methods* section, and your instructor will explain how to use the particular balance(s) available in your lab.

Remember, the balance must always read exactly zero when nothing is on the balance pan. (It is a good idea to check it and tare it before each measurement.) Also, the readings should be recorded to at least the nearest 1/100th of a gram (0.01 g). For this measurement, you will record the digit two places to the right of the decimal point, even if this digit is zero. If your balance shows a third decimal place, record it.

Caution! A single drop of water in the bag will add enough weight to impact the results. Make sure the bag is dry at the start and weigh all of gases before you add water to determine the volume of the bag.

1. Gather materials for the weighing device, including a plastic bag, a one-hole rubber stopper with a medicine dropper stuck through the hole, a cork ring with a ring that fits the rubber stopper, and a dropper cap (see *Figure 9.1*).

2. Push the opened end of the plastic bag through the center of the cork ring. Open the bag and push the rubber stopper firmly into the open bag so that the bag is tightly caught between the stopper and the ring. Make sure there are no leaks in the system. To do this, inflate the bag with one of the gases, put the rubber cap over the hole in the medicine dropper, and gently squeeze the bag. No air should escape.

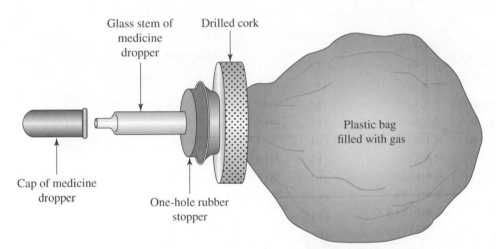

Figure 9.1 Diagram of the assembled system

3. Take the rubber cap off of the medicine dropper and press all air out of the bag. Be sure to completely flatten the bag to get all the air out so that you are weighing a completely empty bag. Once the air is out of the bag, replace the cap on the medicine dropper.

4. Tare the balance so that it reads 0.00 g. Weigh the assembly and record the weight to the nearest 0.01 gram.

5. Remove the cap and connect the bag to a source of gas using the medicine dropper. Hold the bag by the stopper and completely fill the bag with the first gas you are investigating.

7. Do not squeeze the bag, but allow any excess gas to escape so that the gas in the bag will be at the same pressure as the room. Replace the rubber cap on the dropper.

8. Weigh the bag assembly containing the gas (that is at room temperature and normal atmospheric pressure) and record your measurement. Make sure that the bag assembly is completely over the balance pan and does not touch any part of the balance frame or housing.

9. Remove the cap from the dropper and press all of the sample gas out of the bag. Repeat the procedure (steps 5–8) for the other gases available in the lab, including the unknown gas assigned to your team.

Part II. Finding the Volume of the Bag

1. After you have weighed all the available gases, fill the bag with air at room pressure and attach a rubber hose to the medicine dropper (*Figure 9.2*).

2. Fill a large pan or the sink with enough water to cover the mouth of a large inverted bottle of water.

3. Completely fill a large bottle with water. Do not leave any air space in the bottle.

4. Cap the bottle and invert it. Place the inverted bottle in the sink or pan of water so that the mouth is below the surface of the water. Make sure there are no air bubbles in the bottle.

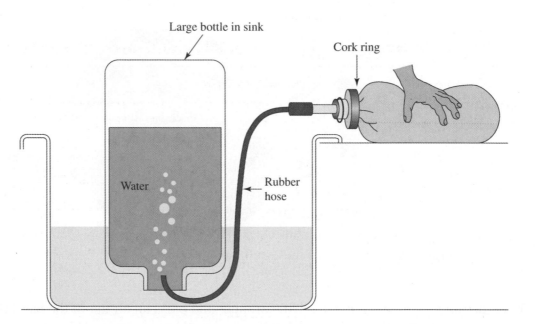

Figure 9.2 Diagram of the setup used to determine the volume of the plastic bag

5. While the bottle opening is underwater, remove the cap from the bottle.

6. With the help of your lab partner, insert the free end of the rubber hose into the mouth of the bottle, as shown in the diagram.

7. Gently squeeze the air out of the plastic bag. As you do so, bubbles of air will enter the bottle and force some of the water out of the bottle.

8. When the bag is empty, keep the bottle inverted with the mouth under water while you replace and tighten the screw cap.

9. Remove the bottle from the sink or pan, place it in an upright position, and remove the cap.

10. Fill a 1,000-mL (1-liter) graduated cylinder with water to exactly the 1,000-mL line. Slowly pour water from the cylinder into the open bottle until the bottle is completely filled. Record the volume of water left in the graduated cylinder. Subtract the volume of water remaining in the graduated cylinder from 1,000 mL and record this number on your data sheet. This value is the volume of water that refills the bottle. Since this amount of water was displaced by the air in the bag, the value also is equal to the volume of air that was originally in the bag.

11. Record the room temperature and air pressure. Your instructor may provide these values for the entire class, or you may be asked to measure them with a thermometer and a barometer.

12. Use Table 9.1 to determine the mass of 1 liter of air at the temperature and air pressure of the room. Record this value on your data sheet. **Note:** In the chart, air pressure is given in "mm Hg," the height of a column of liquid mercury in a mercury barometer. For comparison, standard atmospheric pressure at sea level is 760 mm Hg.

Table 9.1 The Mass (in grams) of 1 Liter of Air at Different Temperatures and Pressures

Pressure	Temperature			
(mm Hg)	15°C	20°C	25°C	30°C
600	0.97	0.95	0.94	0.92
610	0.99	0.97	0.96	0.94
620	1.00	0.98	0.97	0.95
630	1.02	1.00	0.99	0.97
640	1.03	1.01	1.00	0.98
650	1.05	1.03	1.02	1.00
660	1.07	1.05	1.03	1.01
670	1.08	1.06	1.05	1.03
680	1.10	1.08	1.06	1.04
690	1.11	1.09	1.08	1.06
700	1.13	1.11	1.09	1.07
710	1.14	1.12	1.11	1.09
720	1.16	1.14	1.12	1.10
730	1.18	1.16	1.14	1.12
740	1.19	1.17	1.15	1.13
750	1.21	1.19	1.17	1.15
760	1.22	1.20	1.18	1.16

Analyzing Evidence

1. Calculate the apparent mass of gas in the bag by subtracting the weight of the empty bag from the weight of the full bag. (This apparent mass may be less than zero. How can that be?)

2. Calculate the mass of air displaced by the bag of gas by multiplying the volume of the bag (in liters) by the mass of 1 liter of air at room temperature and pressure (taken from Table 7.1 on the previous page).

3. Calculate the corrected mass of the gas in the bag by adding the mass of the displaced air to the apparent mass of the gas.

4. Use the defined molar mass of oxygen (32 grams/mole) to calculate the investigation molar masses of the other gases, by means of the following equation.

$$\frac{\text{molar mass of gas A}}{\text{molar mass of oxygen}} = \frac{x \text{ g/mol}}{32 \text{ g/mol}} = \frac{\text{measured mass of gas A}}{\text{measured mass of oxygen}}$$

5. Record all of these results in the table on your data sheet.

Interpreting Evidence

1. Using the molar masses you calculated from your investigation results, rank the following gases from lowest to highest molar mass: oxygen, methane, carbon dioxide, argon, and nitrogen. Compare this ranking to your predicted ranking. Explain any differences and use evidence to support the ranking you think is correct.

2. Weighing a gas is more complicated than weighing a small solid object. Describe how the following errors in the measurement of mass would change the data. Would your mass be too low or too high? Why?

    a. Not emptying the bag completely before weighing the assembled system
    b. Having the bag resting on part of the balance while measuring the weight of a gas
    c. Not correcting for buoyancy

3. Compare your values for the molar mass of each gas with the accepted value. Do your results support the hypothesis that equal volumes of gases have equal numbers of molecules? Briefly explain the basis for your answer.

4. What are some likely sources of error that could account for the difference between measured and accepted values for molar mass? Indicate whether each one would make your answer too low or too high. (**Example:** How would the results of your investigation change if the temperature of one of the gas samples changed before it could be weighed?)

Making Claims

What can you claim about the molar masses of different gases?

Reflecting on the Investigation

1. One of the first ways that was used to determine the molar mass of a liquid compound utilized the following procedure. The liquid was placed in a glass container, and the container was heated until all of the liquid had boiled away. As a consequence, the air originally in the container was totally displaced by gaseous molecules of the unknown compound. The container was sealed (while still hot), then cooled and weighed. What information do you think was needed to complete the determination of the molar mass of the liquid?

2. Why can't we weigh a single molecule? Explain why finding the mass of a mole of molecules is a good substitute.

Notes

Verifying Molar Ratios in Chemical Reactions

Asking Questions

- What is the relationship between amounts of reagents and amounts of products?
- How do the concepts of moles and molar ratios help define the changes in chemical reactions?
- How can we use the masses of reagents and products to verify the molar ratios in a balanced equation?

Preparing to Investigate

In this investigation, you will collect quantitative measurements for a reaction you examined in investigation 1, specifically the reaction of sodium bicarbonate, $NaHCO_3$, with hydrochloric acid, HCl. Sodium bicarbonate is also known as sodium hydrogen carbonate, but you may be familiar with it as baking soda. The reaction of sodium bicarbonate with hydrochloric acid produces table salt (sodium chloride, $NaCl$), water, and carbon dioxide. The balanced chemical equation is shown here.

$$NaHCO_3 + HCl \rightarrow NaCl + H_2O + CO_2$$

Previously, you used this reaction to make carbon dioxide in a plastic bag. Now you will determine by investigation how many moles of salt ($NaCl$) are formed from a known number of moles of $NaHCO_3$. During chemical reactions, substances combine with each other in a definite proportion by mass, meaning that only a certain amount of one reagent will react with a given amount of another reagent. The amounts of reactant species can be expressed in a variety of ways: grams, pounds, tons, or liters. However, no matter what units are used, they are all related to the ratio of *moles* of one species reacting with a certain number of *moles* of another species. If you are unclear about the concept and definition of a **chemical mole**, you should review the discussion of moles in Chapter 3 of *Chemistry in Context*.

Based on the chemical equation shown above, the molar ratio of reactant to product should be equal to one. This means that for every mole of sodium bicarbonate that reacts with hydrochloric acid, a mole of sodium chloride will be formed. In this investigation, you will verify this ratio by determining the mass of $NaHCO_3$ used and the mass of $NaCl$ formed by weighing samples on a balance. The masses can be converted to moles, and the molar ratio can be calculated. You also can repeat the process with a different reagent and verify the molar ratio for the production of sodium chloride from sodium carbonate (Na_2CO_3). To further improve the accuracy of the results, each group will collect multiple data points for each compound, and the class data will be combined and analyzed.

Making Predictions

- Write out the balanced equations for the chemical reactions between sodium bicarbonate and HCl and sodium carbonate and HCl. Without doing any calculations, predict the molar ratios of reactants and products with each reaction and explain why you think these ratios reflect the chemical reactions.

- After reading *Gathering Evidence*, prepare a data sheet that starts with the equations and predicted ratios. Make a table for collection of data that includes a column for each of the three trials and rows for the mass of the tube, mass of the tube with compound, mass of the compound, molar mass of the compound, moles of the compound, mass of the tube with product, mass of the product, moles of the product, and molar ratio. If you are investigating both sodium bicarbonate and sodium carbonate, make sure you have enough rows to investigate both compounds.

- Using values for atomic weights from the periodic table, calculate the molar mass of each compound (reagents and product) you will measure in the investigation to the nearest whole number. Enter these values into the table on your data sheet as accepted molar masses for the compounds.

Gathering Evidence

Overview of the Investigation

1. Label and weigh three test tubes.
2. Add sodium bicarbonate to each test tube, reweigh the tube, and calculate the mass of sodium bicarbonate in each tube.
3. React the sodium bicarbonate with 10% hydrochloric acid.
4. Evaporate the liquid remaining in the test tube after the reaction takes place.
5. Determine the weight of sodium chloride produced.
6. Calculate the ratio of moles of NaCl formed to moles of $NaHCO_3$ used.
7. Repeat process using sodium carbonate instead of sodium bicarbonate.

Conducting the Reaction and Measuring Reactant and Product Masses

Notes about weighing: This investigation requires careful weighing of test tubes on a laboratory balance. The technique of measuring mass is described in detail in the *Laboratory Methods* section of this book, and your instructor will explain how to use the particular balance(s) available in your lab. Remember, the balance must always read exactly zero (0.00g) when nothing is on the balance pan. It is a good idea to check it and tare it before each measurement

1. Obtain three test tubes that are clean and completely dry. Label them <u>at the top of the tubes</u> with labels A, B, and C. Add a small boiling chip to each test tube.

2. Take the test tubes and your data sheet to one of the laboratory balances. Weigh each of the test tubes (including the boiling chip) to the nearest hundredth of a gram (0.01 g) and record

the masses in your data table. For accuracy, be sure to record the values to the second decimal place, even if the second decimal place is a zero (e.g., 18.10 g).

3. To each test tube add just enough $NaHCO_3$ to fill the curved bottom of the tube.

4. Weigh each test tube again with its contents to the nearest hundredth of a gram (0.01 g) and record the mass in your data table. **Note:** The masses of the three solid samples do not need to be identical. Typical masses of added $NaHCO_3$ will be about 0.30 to 0.70 g.

5. At your lab bench, fill a plastic pipet with 10% hydrochloric acid solution. Add the acid dropwise to tube A. Let each drop of the liquid run down the wall of the test tube and, after each drop reaches the bottom, gently tap the tube. Continue to add acid slowly until all of the solid has dissolved. Keep in mind that it is important to add only the *minimum* amount of acid needed to dissolve the solid. Put the tube aside for Step 7.

 CAUTION! 10% HCl is corrosive to the skin and other materials. Avoid spilling it on yourself, your partner, or your work space.

6. Repeat step 5 with each of the remaining test tubes (B and C).

7. To evaporate the water, you will need to heat the tubes to dryness over a Bunsen burner.

 STOP! No flammable chemicals should be in the vicinity of the Bunsen burner. Long hair should be tied back and extremely loose sleeves on clothing should be avoided.

NOTE: Too rapid heating of the tube, especially if it is held in an upright position, will cause the hot contents to splash out of the tube and will necessitate starting over with a fresh sample. Boiling chips should help to produce smooth boiling. Your instructor may provide additional advice on how to minimize the problem.

As in *Figure 10.1* (next page), hold test-tube A at an angle and point it away from you and anyone else in the immediate vicinity. Gently heat the tube and its contents over a Bunsen burner flame. You want to evaporate the water in the tube without its contents boiling over or splattering.

Continue heating until *all* of the liquid has evaporated and solid NaCl remains. (It is crucial to the success of this investigation to be sure that all of the water has evaporated from the *upper* part of the tube.)

8. Remove the tube from the flame and test for the evolution of water vapor from tube A by inverting a clean, dry test tube over the upright mouth of test tube A. If condensation occurs in the cold test tube, continue the drying and testing process until no condensation occurs. Then set test tube A with its dried contents aside to cool.

9. Repeat the procedures in steps 7 and 8 with test tubes B and C.

10. Allow the three test tubes to cool (at least 5 minutes), *check to be sure there are no water droplets left*, and then weigh each with its contents. Record the masses in the data table.

11. If time permits, confirm that the tubes were fully dried by reheating them for 1–2 minutes, cooling for 5 minutes and reweighing. Record the second mass for each. If they were dry the first time, there should be a negligible change in mass.

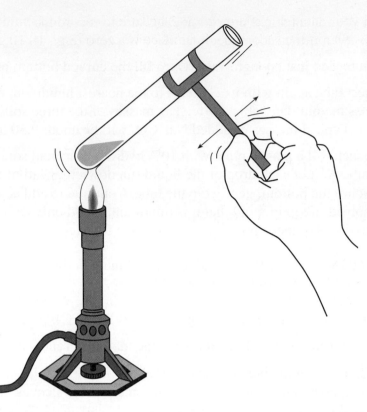

Figure 10.1 The correct way to heat a test tube over a Bunsen burner

12. After completing the study with sodium bicarbonate, you may try a similar study with sodium carbonate (Na_2CO_3) to see how the results differ. If so, repeat Steps 1-10 with clean, dry test tubes. If you do this version, you will need to have additional lines on your data table for the masses and calculations.

Clean-up

Dispose of the test tube contents as your instructor directs you. Do not put anything down the sink drain without being told that is permitted. Thoroughly wash the test tubes, rinse with deionized water, and leave them upside down to drain.

Analyzing Evidence

1. For each test tube, subtract the mass of the test tube from the test tube with added $NaHCO_3$ to determine the mass of $NaHCO_3$ used.

2. Use the molar mass of $NaHCO_3$ that you previously calculated and the masses from Step 1 to calculate the number of moles of $NaHCO_3$ used in each trial.

3. For each test tube, subtract the mass of the tube from the final mass of the test tube with product to obtain the mass of $NaCl$ formed by the reaction.

4. Proceeding the same way as in step 2, calculate the moles of $NaCl$ formed.

5. Calculate the <u>ratio</u> of moles NaCl to moles $NaHCO_3$. This ratio should be recorded to two decimal places (i.e., 2 digits after the decimal point).

6. Calculate the average of your three investigation results for this ratio and record it to two decimal places.

7. Repeat calculations for $NaHCO_3$.

Interpreting Evidence

1. Combine your data with those of the rest of the class. Do your investigation results, or those of the class, agree with the balanced equation? Discuss your answer. If you think any results should be excluded from calculating the average class ratio, explain why and calculate a revised average ratio.

2. Suggest two possible sources of error in this investigation (do not include weighing errors). Indicate whether each error would increase or decrease the investigation value for the mole ratio.

Making Claims

What can you claim about chemical equations and molar ratios of reactants and products? What evidence from this investigation can you use to defend your claim?

Reflecting on the Investigation

Balance each equation in the mechanism showing formation of ozone (O_3) in the upper atmosphere. Use this process to determine how many moles of O_2 are required to form one mole of O_3. Defend your answer using evidence from the investigation of the reaction of sodium bicarbonate and HCl.

a. First, nitrogen and oxygen gas react to form nitrogen oxide.

$$N_2 + O_2 \rightarrow NO$$

b. Then, nitrogen oxide reacts with more oxygen to form nitrogen dioxide.

$$NO + O_2 \rightarrow NO_2$$

c. When nitrogen dioxide is struck by UV light, it decomposes to form nitrogen oxide and a highly reactive oxygen atom.

$$NO_2 + UV\ light \rightarrow NO + O$$

d. Finally, the oxygen atom reacts with diatomic oxygen to form ozone.

$$O + O_2 \rightarrow O_3$$

Notes

Hot Stuff: An Energy Conservation Problem

Asking Questions

- What are some ways to measure what you can't see?
- How do scientists approach problems that don't have obvious solutions?
- What data is important when defending a method for solving a problem?

Preparing to Investigate

This laboratory investigation is a departure from the usual investigation. It simulates the kind of problem solving that takes place in a scientific laboratory. There are no instructions, procedures, or data sheets. There is simply a problem to solve that requires reasoning skills and the application of previous experience or knowledge.

A variety of materials and equipment will be available that you can use to solve this problem. At a minimum your "lab" should have graduated cylinders, burets, beakers, flasks, plasticware, Styrofoam cups, stirrers, test tubes, plastic pipets, a balance, and, of course, a 40°C thermometer.

Gathering Evidence

Organizing the Team

Your team must devise a method for solving the problem and then obtain a numerical answer. A student team will typically consist of three or four members. Although all members of the team should be involved in the entire process, your investigation will be more efficient if you take a few minutes at the beginning to be sure each person has a clear role to play. Possible positions you might want to assign to members of the group include a team leader, a "go-fer," an experimentalist, a report writer, and a recorder. It especially will be important to keep track of ideas, data, and investigation procedures during the investigation. When finished, your instructor will lead a class discussion where teams will compare answers and methods.

Measuring Water Temperature

As an energy conservation method, you decide to turn down your home water heater so that it only heats water to 55°C. (It previously heated the water to between 60°C and 70°C.) Unfortunately, when you decide to measure the temperature of the water in your water heater, the only thermometer available has a maximum temperature of 40°C.

In the laboratory, there is a simulated water heater, consisting of a large coffee pot filled with hot water. Each team will have available a thermometer that has the graduations above 40°C covered so that you cannot see them. Using your 40°C thermometer and materials available to you in the lab, devise a way to measure the temperature of the water in the coffee pot.

Optional: Your instructor may ask you to propose two or three *different* methods for solving this problem, then test each of them, decide which gives the most reliable answer, and explain why.

Analyzing and Interpreting Evidence

Your team should be prepared to defend your answer and the method you used to obtain it. Therefore, it is important for your team to keep a complete record of everything you do and the numerical data you obtain. It also is important that you present the data to provide evidence that your method solves the presented problem. In addition to an oral report to the class, your group should prepare a brief (one-page) written report of your investigation. Your instructor will specify what should be included in the report

Making Claims

What can you claim about your method for measuring the temperature of the hot water? Include these claims in your oral and written reports.

Reflecting on the Investigation

When everyone is finished, your class will assemble to hear a report from each student team about the method used and the results. Take notes and identify the strengths and weaknesses of each team's approach.

Comparing the Energy Content of Fuels

Asking Questions

- How can the energy content of fuels be measured?
- Are some fuels more efficient than others?
- Which fuel do you think would release the most energy upon burning, ethane (C_2H_6) or ethanol (C_2H_6O)? Why?
- What is the relationship between the presence of oxygen in a fuel and its energy content?
- Why can't we use carbon dioxide or water as fuels? What prevents argon from being a fuel?

Preparing to Investigate

In this investigation, you will investigate the energy content of several fuels by using them to heat water. The data that you and your classmates obtain will enable you to compare fuels to see which ones provide more energy for a given mass of fuel burned. In particular, you will be able to determine how the energy released by burning **hydrocarbons**, compounds that contain only hydrogen and carbon, compares to that released by burning oxygenated fuels such as ethanol, a renewable **biofuel** derived from biological rather than geological sources.

Fuels burn to give off energy. They do so by combining with oxygen to form compounds of lower energy, typically carbon dioxide and water. For example, here is a chemical equation that represents the **combustion** of methane, CH_4, to produce CO_2 and H_2O.

$$CH_4 + 2\,O_2 \longrightarrow CO_2 + 2\,H_2O$$

Methane burns cleanly. We call this *complete combustion* because the two products, CO_2 and H_2O, are not flammable, that is, they cannot burn any further. Other hydrocarbons, such as propane (C_3H_8), which is used in barbecue grills, and butane (C_4H_{10}), which is used in camping stoves, will burn to produce the same products: CO_2 and H_2O.

Other fuels, however, may burn incompletely unless supplied with plenty of oxygen. In addition to carbon dioxide, the combustion products also include carbon monoxide and/or soot. For example, some hydrocarbons, including candles, burn with a sooty flame. When either soot or carbon monoxide or both are formed as products, this is called *incomplete combustion*. Here is the chemical reaction in which hexane (a hydrocarbon, C_6H_{14}) burns to produce carbon monoxide.

$$C_6H_{14} + 2\,O_2 \longrightarrow 6\,CO + 7\,H_2O$$

Some fuels such as ethanol (ethyl alcohol, C_2H_5OH) contain oxygen in addition to carbon and hydrogen. In essence, they are hydrocarbons that have already partially reacted with oxygen:

$$2\,C_2H_6 + O_2 \longrightarrow 2\,C_2H_5OH$$

The energy when fuels are burned is given off primarily in the form of heat, with some light as well. Although sometimes quite bright, the light emitted by a flame or fire is hard to quantify. In contrast, it is straightforward to measure the heat released when a fuel burns. For these two reasons, you will evaluate the energy content of fuels in this investigation by the heat they give off rather than by the light. So how do you measure the heat? Rather than doing this directly, you will measure the heat released to increase the temperature of a known quantity of water. From this, you can estimate the heat released by burning a particular fuel.

Some of the fuels you will test will be alcohols such as methanol (CH_3OH), ethanol (C_2H_5OH), isopropanol (C_3H_7OH), or butanol (C_4H_9OH). Others will be hydrocarbons, such as lamp oil or candle wax. The latter are actually mixtures of hydrocarbons, but we will approximate their compositions as $C_{12}H_{26}$ (lamp oil) and $C_{40}H_{82}$ (candle wax).

You will heat water by burning a known amount of a fuel. It takes exactly 1 calorie (cal) of heat to raise the temperature of 1 gram of water by 1°C. Therefore, if you know the mass of water and how many degrees the temperature increases, then the total amount of heat absorbed by the water can be calculated:

$$\text{heat absorbed (calories)} = \text{mass of water (grams)} \times \text{temp. change (°C)} \times 1.00 \text{ cal/g•°C}$$

$$= m \times \Delta T \times 1.00 \text{ cal/g•°C}$$

In this equation, ΔT represents "change of temperature", and the last term, 1.00 cal/g•°C, is the **specific heat** of water. Convince yourself that combining and canceling the units on the right side will leave only calories. Theoretically, the amount of heat liberated by the burning fuel should equal the heat absorbed by the water, but in practice, some of the heat will be lost to the surroundings.

Making Predictions

- List the fuels you will study in this investigation and predict their relative energy outputs.

- After reading *Gathering Evidence* and the appropriate parts of *Laboratory Methods*, prepare a data table to record your masses, temperatures, and heat absorption results.

Gathering Evidence

Overview of the Investigation

1. Assemble the apparatus and obtain a burner containing a known fuel.
2. Add a measured volume of water to the can and determine the mass of the water.
3. Weigh the burner with the fuel.
4. Record the initial temperature of the water.
5. Light the burner and heat the water until the temperature increases about 20° C.
6. Extinguish the burner and record the highest temperature of the water.

7. Weigh the burner again to find the mass of fuel used.
8. Repeat with two or more additional trials.
9. For each trial, calculate the amount of heat released per gram of fuel burned.

Part I. Measuring the Energy Content of Different Fuels

The detailed procedure for **calorimetry** is described in the *Laboratory Methods* section of this lab manual, and you should read that section before proceeding.

Note: The success of this investigation depends heavily on the accuracy of your mass measurements. Perform all weighing operations on a laboratory balance. You should read the section on measuring mass in *Laboratory Methods,* and your instructor will explain the use of the particular balances in your lab. For each measurement, it is important to know that the balance reads *zero* with nothing on the balance pan. For some laboratory balances, the "zero" can be changed easily and perhaps unintentionally, causing your measurements to be in error. It is important to check and, if necessary, reset the "zero" *each* time you make a measurement.

Your instructor will tell you which fuel or fuels you are to investigate. You may be asked to do one trial with each of three different fuels. Alternatively, you may be asked to do several trials with a single fuel.

 CAUTION! You need to be constantly aware that you (and other students) are working with flammable solvents and open flames. Handle the fuel burners very carefully. Before starting the investigation, be sure you know where a fire extinguisher is located in your laboratory. As with any investigation involving open flames, long hair must be tied back and loose sleeves should be avoided. Wear eye protection at all times.

Part II. Optional Extensions

1. **Investigate the effects of changes in procedure**. These might include (a) adding some nonflammable insulation around the can, (b) adding a cylindrical shield of aluminum foil around the burner, (c) using 200 mL of water and only a 10°C temperature rise (or 50 mL of water and a 40°C temperature change). Predict what will happen with each change and evaluate your hypothesis based upon your observations.

2. **Energy content of wood**. Design and carry out an investigation to measure the energy content of wood. (A suggested source of wood is the wood splints that are commonly available in chemistry laboratories.) First identify the investigation challenges, and devise a way to solve them.

3. **Caloric content of nuts**. Using nuts as an example of a high-calorie food, design and carry out an investigation to measure the energy content in nuts and compare it with other fuels. First identify the investigation challenges, and then devise a way to solve them.

Clean-up

Return your burners and other equipment to their proper place.

Analyzing Evidence

1. Calculate the temperature change by subtracting the initial temperature from the final (highest) temperature.

2. Calculate the mass of fuel burned by subtracting the final weight of the burner from the original weight of the burner.

3. Do the following calculations, using the equation given in the introduction and the measurements you have recorded on your data sheet.

    a. For each trial, calculate the total calories of heat absorbed by the water. This will be assumed to be equal to the amount of heat liberated by the burning fuel.

    b. Then calculate the calories of heat per 1 gram of fuel burned.

    c. **Optional:** You may be asked to calibrate your measurements using methanol.

4. Before doing another measurement, take a few moments to discuss the procedure and calculations with your partner. Did you encounter any difficulties? Can you think of any desirable improvements in the procedure? Do the calculations for each trial as soon as it is finished to see how the results are coming out.

Interpreting Evidence

Although it was only possible for you to do a few trials, it is desirable to assemble a much larger body of data from your whole class or lab section so that more reliable comparisons of fuels can be made. Your instructor will tell you how to post or report your results for the rest of the class to see. You may be asked to calculate class averages for each fuel that was used. These questions should help to focus your interpretation of the results.

1. Rank the fuels according to the heat per gram that you and your classmates calculated. Identify each fuel in your ranking as either a hydrocarbon or alcohol. What general rule can you propose about the heat content of hydrocarbons compared to that of alcohols?

2. Which class of fuels, hydrocarbons or alcohols, contains oxygen? Based on your rankings, how does the presence of oxygen in the structure of a fuel affect its heat content? Give a reasonable explanation for the difference.

3. Were your predictions of relative energy outputs correct? Explain any differences between your predictions and your results.

4. There are many possible sources of error in this investigation. List three that you can think of. Would each error have a *large* effect, a *medium* effect, or a *small* effect on the calculated heat content of a fuel? Also indicate whether the calculated result would be *too high*, *too low*, or could go *either way*.

Making Claims

What can you claim about the energy content of hydrocarbon and alcohol fuels? What variables in the structure of the fuel affect the energy released from the fuel when it is burned?

Reflecting on the Investigation

1. Gasohol is a mixture of gasoline (hydrocarbons) and ethanol (an alcohol).

   a. How would the energy content per gram of gasohol compare to that of plain gasoline?

   b. How do you think this would affect a car's fuel efficiency (miles per gallon)?

   c. Consider that gasoline comes from petroleum, while ethanol can be made from corn.

      • Explain why environmentalists might promote the use of gasohol.

      • Explain why farmers and farm advocates might promote the use of gasohol.

      • Explain how someone concerned with feeding the poor might feel about increasing use of gasohol.

   d. Because of concerns about the effects of producing fuel ethanol from corn on global food prices, there is currently a lot of research into alternative raw materials for ethanol production. Go online and search for information about other natural resources that can be used for making ethanol. List a few of these materials and explain some of their advantages. Be sure to cite your sources.

2. Carbohydrates have the general formula CH_2O, while fats typically have a general formula of about $C_{10}H_{19}O$. Which of these classes of foods has the greater oxygen content as a percentage of the molecule's mass? Since foods are simply fuels for the body, which of these will release more energy when metabolized in the body? Explain your reasoning. How does this conclusion compare to the relative calorie content of fats and carbohydrates? (If you are not sure how many dietary Calories are in a gram of fat or carbohydrate, see Chapter 11 in your textbook.)

3. Suppose you put 50 mL of water in the can instead of 100 mL and heated it 40 degrees instead of 20 degrees. In what ways, if any, would this affect the results of the investigation? What would happen if you used 200 mL of water and heated it only 10 degreees? Can you think of any advantages or disadvantages in using either 50 mL or 200 mL of water in this investigation?

Notes

Preparation and Properties of Biodiesel

Asking Questions

- Compare biodiesel derived from cooking oil with diesel fuel derived from crude oil. Make your comparison on the basis of (1) the chemical composition, (2) their source in the environment, and (3) the process by which they reach your fuel tank.
- Does it require more energy to prepare biodiesel from waste cooking oil than is returned when the fuel burns?
- How might use of used cooking oil and discarded animal fats complicate the production of biodiesel?
- How might the use of used cooking oil and discarded animal fats in the production of biodiesel lead to more sustainable fuel sources?
- How can new research opportunities, such as turning glycerol into biodegradable plastics, improve the economics and environmental impacts of biodiesel?
- What are some of the potential problems if a fuel gels at typical winter temperatures?

Preparing to Investigate

Waste oils from cooking can be burned directly for heat, but they don't have the right characteristics to substitute for gasoline or diesel fuel in cars. Fortunately, a simple chemical reaction converts oils to a more useful form called **biodiesel**. Biodiesel can be produced from waste or new fats and oils, but waste cooking oil contains impurities that complicate the biodiesel synthesis. At the discretion of your instructor, you will use either waste cooking oil or new cooking oil off the supermarket shelf to synthesize biodiesel during this investigation.

To synthesize biodiesel (see *Figure 13.1*), triglycerides in oils and fats react with methanol in the presence of a catalyst (NaOH) to form two products. Triglycerides contain three ester functional groups (see sections 4.10 and 11.3 in *Chemistry in Context*). They are even larger molecules than they appear in *Figure 13.1* because each R-group represents a long hydrocarbon chain containing

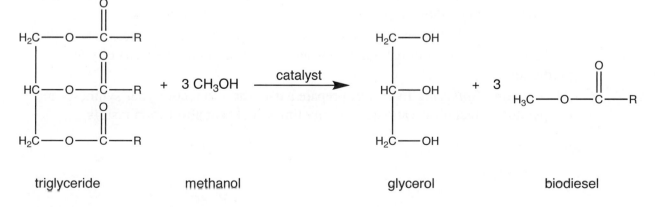

triglyceride   methanol   glycerol   biodiesel

Figure 13.1 Synthesis of Biodiesel

10-20 carbon atoms. The first product is glycerol, a "triple" alcohol, also called glycerin. Glycerol is a syrupy liquid that is useful for making cosmetics and other consumer products. Increases in biodiesel production have led to a glut of glycerol on the market, so green chemists are now seeking new uses for it. The second product is biodiesel, a methyl ester of one of the three long chain fatty acids that initially was part of the triglyceride.

Since vegetable oils are mixtures of several different triglycerides, biodiesel fuels will also be mixtures of several molecules that differ in the number of carbon atoms in the R-group. Each type of vegetable oil produces a different mixture of biodiesel molecules. As a result, biodiesel fuels prepared from different types of oils or fats have somewhat different properties. In this investigation, you will examine three properties of the biodiesel you prepare: viscosity, gelation temperature, and energy content. All are important to its use as a fuel.

Viscosity is a measure of how easily a liquid flows or pours. For instance, water has a low viscosity and flows easily, whereas maple syrup has a much higher viscosity and pours slowly. Because a fuel must flow through the engine, its viscosity is important to the proper functioning of that engine. You will measure the viscosity of both your biodiesel product and the oil from which it was prepared by measuring how long it takes each liquid to drip from a small pipet. The longer it takes for the liquid to drain, the higher the viscosity of the liquid.

Diesel fuel thickens and turns to a gel in cold weather. Diesel engines are notoriously difficult to start at low temperatures because gelled fuel will not flow or ignite properly. You will observe the biodiesel and oil to see if they cloud and gel at winter temperatures near 0°C.

Of course, the most important characteristic of a fuel is that it burns to release energy. If you did Investigation 10, you have already measured the energy content of several fuels. You will follow that same procedure to measure the energy content of your biodiesel. Consult Investigation 10 for additional information.

Making Predictions

- Biodiesel and a glycerol-water mixture are not soluble in each other. Which floats on top of the other? Explain the basis of your prediction.
- When a liquid flows, its molecules get tangled up. More tangling makes the liquid flow more slowly. Predict whether the viscosity of the oil will increase or decrease upon reaction to form biodiesel. Explain the basis for your prediction.
- Biodiesel is a fuel. Molecule per molecule, does it produce more or less energy than methane (a component of natural gas) when burned? Explain the basis of your prediction.
- After reading *Gathering Evidence*, prepare a data sheet to record your synthesis and temperature effect observations, viscosity times, and heat absorption results.

Gathering Evidence

Overview of the Investigation

1. Prepare a mixture of methanol and NaOH.

2. Heat the oil (waste or new) to 50° C.

3. Add the methanol mixture to the oil and stir for 20 minutes.

4. Centrifuge the product and then remove top biodiesel layer for testing.

5. Measure the viscosity of the oil and biodiesel.

6. Observe the effect of low temperatures on the oil and biodiesel.

7. Add a wick and some biodiesel to a 20 mL beaker.

8. Follow the calorimetry procedure to measure the energy content of the biodiesel.

Part I. Preparing Biodiesel

1. If your class is preparing biodiesel from several types of oil, your instructor will tell you which to use. Record the type of oil.

2. Use a 10 mL graduated cylinder to measure 10 mL of methanol into a small beaker.

3. Add 0.5 mL (about 10 drops) of 9M NaOH to the methanol. Record observations. Set this mixture aside until step 7.

   CAUTION! NaOH is caustic and can harm skin. If you spill any on your hands, wash them immediately and notify your instructor.

4. Use a 50 or 100 mL graduated cylinder to measure out 50 mL of oil. Pour the oil into a 250 mL beaker.

5. Add a magnetic stirring bar to your oil and place the beaker on a stirring hotplate. Adjust the stirring rate so that the solution is being well mixed without splashing. Turn the heat on *low*. Alternatively, your instructor may give you a different procedure where you will shake, rather than stir, your reaction.

6. Heat your mixture to 50° C. To test the mixture's temperature, hold a thermometer in the mixture but well away from the spinning stir bar. Do not stand the thermometer in the beaker because it will break if struck by the stir bar.

7. As soon as your oil reaches 50°C, *turn off the heat*, but keep the oil stirring. Slowly pour your methanol/NaOH mixture into the oil. Be sure that the mixture continues to stir sufficiently so that the methanol does not form a layer on top. Record observations.

8. After 20 minutes, remove your beaker from the hotplate. Record observations.

9. Obtain four centrifuge tubes. Fill each three-quarters full with your biodiesel mixture. Place the tubes opposite to each other in a centrifuge and spin them for several minutes.

10. Turn off the centrifuge and allow it to stop before trying to remove your tubes. You should see two well-separated layers. If not, return the tubes to the centrifuge for several more minutes. The top layer is your biodiesel. The bottom layer is glycerol.

11. Use a disposable pipet to transfer the top biodiesel layers to a small beaker. You will need about 20 mL of biodiesel in total. If you don't have this much, pour more of the reaction mixture from the 250 mL beaker into centrifuge tubes and repeat the separation process.

12. Keep the small beaker of biodiesel for testing. Discard the lower layer of glycerol and any of the mixture remaining in the 250 mL beaker as directed by your instructor.

Part II. Determining the Viscosity of Oil and Biodiesel

1. Obtain two 6-inch disposable glass pipets that have been marked with lines on the barrels 4 cm apart, as shown in *Figure 13.2*. You'll also need a matching dropper bulb and a timer that can measure in seconds.

2. We will estimate viscosity by measuring the time required for the liquid level in the pipet to drop from the upper line to the lower line. Viscosity is proportional to this time.

3. Use the bulb to fill one of your pipets with your biodiesel so that the liquid is above the upper line. Hold the pipet vertically over the beaker of biodiesel. Quickly remove the bulb and place your index finger firmly over the top of the pipet. This will stop the biodiesel from dripping.

4. Have your partner ready to time. Remove your finger from the top of the pipet. Allow the liquid to drain freely back into the beaker. Measure the time required for the biodiesel level to drop from the upper line to the lower line. Record your time.

5. Repeat the measurement two more times. Record your data and discard the used pipet.

6. Transfer about 5 mL of the oil used in Part I as the starting material for your biodiesel to a test tube. Use your second marked pipet to measure the viscosity of the oil. Allow the oil to drip back into its test tube. Do the measurement three times and record your data. Save your oil for measuring the heat content in Part III.

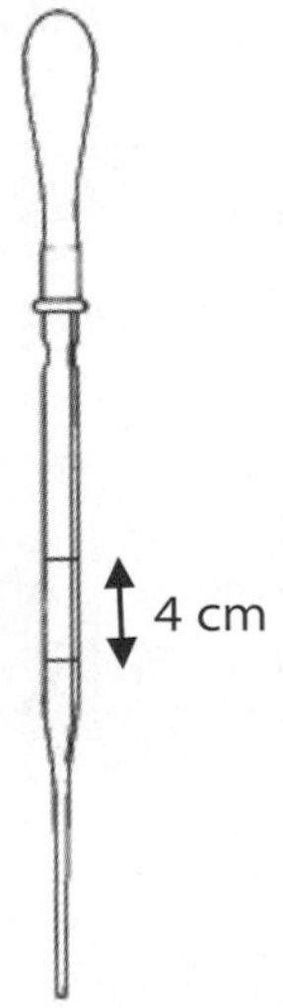

Figure 13.2 Pipet for viscosity measurement

Part III. Observing the Effect of Temperature on Oil and Biodiesel

1. Fill a 250 mL beaker most of the way with ice. Add two large scoops (at least 2 teaspoons full) of salt to the ice and then add enough water to make the mixture slushy. Stir briefly to mix in the salt. The presence of salt lowers the temperature of the ice bath.

2. Place a thermometer inside your test tube of oil. Note how cloudy or clear the oil appears. Lift the thermometer briefly and note how viscous (thick) the liquid appears as it drips from the thermometer or flows down the inside wall of the test tube. Place the test tube of oil in your ice bath.

3. Observe the oil as it cools. Periodically remove it from the ice long enough to see if the oil is becoming cloudy or more viscous. If so, record the oil's temperature. Keep the test tube in the ice until its temperature is no longer dropping. Again note the oil's appearance and apparent viscosity. Record your observations.

4. Wipe the oil from your thermometer with a paper towel or tissue.

5. Put some of your biodiesel in a test tube, add the thermometer, and repeat the cooling and observation process.

Clean-up

Discard or recycle the contents of your test tubes as directed by your instructor. Clean your thermometer with soap and water to remove all traces of the oil.

Part IV. Measuring the Heat Content of Biodiesel

 STOP! You need to be aware constantly that you (and other students) are working with flammable solvents and open flames. Handle the fuel burners very carefully. Before starting the investigation, be sure you know where a fire extinguisher is located in your laboratory. As with any investigation involving open flames, long hair must be tied back and extremely loose sleeves should be avoided. Wear eye protection at all times.

1. Prepare a burner for your biodiesel. Obtain a 20 mL beaker, a candle wick, and a paper clip. The end of the candle wick should be wrapped around one wire of the paper clip. Spread the paper clip out slightly so that it makes a stable bottom support for the wick. The wire inside the wick will allow it to stand vertically. Put the paper clip and wick in your beaker. The wick should extend to the top of the beaker. Now fill the beaker with at least 15 mL of your biodiesel fuel.

2. The wick should stand freely in the middle of the biodiesel. Use matches to light the wick and ensure that your burner is working well. Once you know that your burner is working, extinguish its flame.

3. Measure the energy content of your biodiesel by following the calorimetry procedure given in the *Laboratory Methods* section. Your apparatus will look similar to that in Figure 10 of that section, except that the burner in this case is your beaker filled with biodiesel. Repeat the measurement twice.

Analyzing Evidence

1. Look carefully at the data from each of the sets of viscosity estimates to see whether the results show consistency or whether any one result in a given set should be eliminated because it appears to be an outlier. If so, draw a single line through the result in your data table and make a note that you have not included it in your analysis.

2. Calculate the average amount of time that it took for each liquid to drain.

3. Calculate the heat released by your biodiesel for each trial and average the results of your trials together.

Interpreting Evidence

Although it was only possible for you to do a few trials, it is desirable to assemble a much larger body of data from your whole class or lab section so that more reliable comparisons of biodiesel fuels can be made. Your instructor will tell you how to post or report your results for the rest of the class to see. You may be asked to calculate class averages for each type of biodiesel if different oils were used. The following questions should help to focus your interpretation of the results.

1. You made a prediction as to which layer in your centrifuge tube would be on top. Which layer actually was on top? Was your prediction correct? If yes, what basis did you use for the prediction? If no, what error did you make?

2. To get an accurate measure of the energy content of the biodiesel fuel, all the heat released when it burns must be transferred to the water in the can. We never get it all. Where else can it go? As a result of these losses, do you expect your calculated heat content to be a little too low or a little too high when compared to the actual heat content of the biodiesel? Explain.

3. If your class prepared biodiesel from several different oils, prepare a small table listing the drip time and the heat content for each type of biodiesel. Now rank the fuels in order from lowest to highest energy content. Compare your ordered list to *Figure 11.9* in the *Chemistry in Context* textbook. What do you notice?

Making Claims

What can you claim about biodiesel as an energy source?

Reflecting on the Investigation

1. You noted in this investigation that biodiesel and water do not mix. Give two other examples of some type of oil not mixing with water.

2. Predict the density of gasoline, given what you know about the density of water and given what you observed in this investigation. Then look up the value to see how close you came.

Detecting Ions in Solution

Asking Questions

- What factors are important to consider when building a scientific instrument?
- How can you determine what solutions contain ions?
- What is the relationship between ion concentration and electrical conductivity?
- Why is pure water often referred to as deionized water?

Preparing to Investigate

In this investigation, you will build a detector and use it to determine whether different solutions contain separate electrically charged particles called **ions.**[1] (See Chapter 5 in *Chemistry in Context* for more detail.) Ionic solutions contain both positively charged and negatively charged ions. For example, when solid sodium chloride, NaCl, is dissolved in water, it breaks apart into Na+ and Cl- ions. These ions carry electric charges and are free to move independently. Therefore, they conduct electricity through the solution. This property provides a very simple and useful way to test for the presence of ions. If ions are present, the solution will conduct electricity. Also, the magnitude of the conductance is directly proportional to the concentration of ionic substances dissolved in the liquid.

Your detector will measure the presence of ions indirectly through determining whether the solution can conduct electricity. Specifically, the light-emitting diode (LED) in your detector will light up when the probes are immersed in a liquid that conducts electricity. This results because conducting materials complete the circuit in your detector, and the current flow causes the solid state LED to emit visible light.

Making Predictions

- Which of the following do you expect to conduct electricity: pure water, 1% salt solution, 1% sodium hydroxide solution, 1% HCl solution, 1% sugar solution, or 10% ethanol solution? Explain why you predict these results.

- After reading *Gathering Evidence*, prepare a data sheet that has room for your predictions and includes a data table that lists the solutions and has blanks to record the conductivity measurements and observations.

---

[1] This design for a conductivity detector was originated by F. J. Gadeck, *J. Chem. Educ.*, **64**, 628 (1987).

Gathering Evidence

Overview of the Investigation

1. Collect the materials for building the detector.

2. Practice soldering, if necessary.

3. Assemble the detector.

4. Test the detector on a known solution.

5. Test various solutions in the laboratory.

6. Test materials outside of the laboratory.

Part I. Soldering Wires

Solder is a low-melting mixture of several metals that is melted onto wires to join them and make an electrical connection. For this investigation, you will either use an electrically heated soldering iron or "gun" or small strips of tape solder that are wrapped around the wires and then heated with a match or candle flame. The tape solder melts at a low temperature and makes a soldering iron unnecessary.

Before assembling the detector, you should first practice soldering together some scrap pieces of wire as follows:

1. Using a wire stripper, strip off about 1 cm of the plastic coating from the ends of two wires.

2. Twist the bare ends of the wires together to make a mechanical connection.

3. Do ONE of the following:

   a. If you are using a soldering iron or gun, obtain a strip of solder wire. Turn on the soldering iron or gun and hold it against the end of the solder wire until the solder begins to melt. This will confirm that the soldering iron is hot. Then touch the hot tip of the soldering iron to the wires to be joined and hold the end of the solder wire to the heated spot. The solder wire should melt and flow smoothly over the hot wire junction. Remove the soldering iron and wait for the joint to cool. When it is cool, pull gently on the ends of the wire to be sure that the junction is formed and that the wires will not come apart.

   b. If you are using tape solder, cut a piece that is 0.5–1.0 cm long. Wrap the tape solder around the connected wires and then heat the tape solder gently with a match or candle. The tape solder should melt and flow over the wires and join them together. When the wires are cool, pull gently on the ends of the wire to be sure that the junction is formed and that the wires will not come apart.

4. Repeat steps 1–3 until you can make a simple solder joint that seems solid.

Part II. Assembling the Detector

1. Gather the parts for the conductivity detector and be sure you can correctly identify each part listed in the caption to *Figure 14.1*. You will also need two 12-inch lengths of thin plastic tubing and a 12-inch piece of black electrical tape.

2. Prepare the film canister cap by punching four holes in it as shown below. Punch the holes from the inside of the cap to avoid damaging the lip of the cap. Use a hole puncher if available. If this is not possible, hold a nail in a flame with pliers until the nail is hot (not red hot) and then use the nail to melt suitable holes in the top of the plastic film container.

3. Assemble your detector as shown in *Figure 14.1*. Note that the letters (A through G) refer to components of the detector that are labeled in the figure. Push the two wires from the LED (part F) through the two closely spaced holes in the cap of the film canister (D).

4. Locate the longer wire of the LED. If thin plastic tubing is available, cut a length so that it covers all of the wire except for about 1 cm. This tubing, sometimes called "spaghetti", insulates the bare wires.

5. Twist the **long** wire of the LED and a wire from the resistor (E) together. (**Note:** If the wrong wire is attached to the resistor, the LED will be permanently damaged when current flows through the circuit.) Solder the resistor and the LED together at this joint.

6. If it is available, cut a length of thin tubing so that it covers all of the other resistor wire except for about 1 cm.

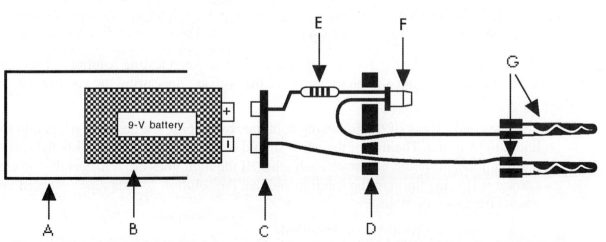

Figure 14.1 Diagram of the finished conductivity detector: (A) film canister, (B) 9-V battery, (C) battery clip, (D) canister cap, (E) resistor, (F) LED, and (G) alligator clips

7. Solder the remaining resistor wire to the red wire from the battery clip (C).

8. Thread a wire through one of the remaining holes in the cap. Locate the other LED wire. If thin tubing is available, cut a length of it so that it covers all of the LED wire except for about 1 cm.

9. Solder the shorter wire of the LED to this wire.

10. Thread a second wire through the cap. Locate the black wire from the battery clip and solder it to the wire.

11. Attach alligator clips (G) to the ends of the two wires extending through the cap. First make a mechanical connection for each; then solder the connections.

12. Test your detector by connecting the battery clip to a 9-volt battery and touching the wires or the alligator clips together. Completion of the circuit should cause the LED to light up.

13. If the detector does not work, check the entire system against the diagram in *Figure 12.2* to see if you have made a soldering error. Also check for loose connections. If you cannot find an error, check with your instructor.

14. If the detector works, package it in a black plastic film canister (*Figure 12.2*). Be careful not to pull apart any connections or have any bare wire connection touching any other bare connection.

Part III. Testing for Ions

If alligator clips are used, do not insert the metallic clips directly into the solutions. Instead, use them to hold short pieces of wire (e.g., a partially unfolded paper clip). The wires can then be inserted into the solutions to be tested. Using wires to test the solutions keeps the alligator clips out of the solutions and prevents the clips from becoming corroded.

The tests can be performed in small beakers, small test tubes, or in a plastic wellplate. Be sure to record your observations on the data sheet as you proceed. For each test, record whether you observe a bright light, a dim light, or no light.

NOTE: It is important to rinse the wires with pure water between testing solutions. This will prevent contamination of your samples and ensure that you are measuring the conductivity of the current sample.

1. Test the conductivity of the following solutions by dipping the attached wires into the liquid to be tested. The tips of the wires should be about half an inch below the surface of the solution and should not touch each other. If the light goes on, it means that an electric current is flowing through the solution and that the solution contains ions. Record your results for these solutions.

    a. Pure water (distilled or deionized)

    b. Tap water

    c. Salt water (1% solution of sodium chloride).

2. Use the sample (pure water, tap water or salt water) that had the brightest light to test whether different lengths of wire that are immersed in the liquid influence your conductivity results.

3. Use the same sample to determine whether the distance between the wires that are immersed in a liquid affects the sensitivity of your detector.

4. Next, test the following samples and record whether you see a bright light, a dim light, or no light from the LED. **NOTE:** If wire length or distance influenced the results, be sure to keep these variables constant when testing new samples.
    a. Sodium hydroxide (NaOH) in water - 1% solution
    b. Hydrochloric acid (water solution of HCl) - 1% solution
    c. Sugar in water - 1% solution
    d. Ethanol (ethyl alcohol – C_2H_5OH) in water - 10% solution

Part IV. Optional Activities

1. Your instructor may provide additional solutions to test in the laboratory. If so, be sure to record what the solutions are and what you observe.

2. Devise and carry out an investigation to determine how sensitive your detector is to the presence of ions in solution. Record your method and the results.

3. Now that you know how to use the conductivity tester, take it with you when you leave the lab and test at least five liquids that you encounter during the course of a day. Enter the data in your data sheet. Be prepared to report your observations at the next meeting of your class or lab section. Some suggestions for liquids to try include (a) tap water from different sources, (b) stream or lake water, (c) rainwater, (d) sweat, (e) urine, (f) foods such as coffee, tea, fruit juice, milk, soft drinks, a potato, a piece of apple or other fruit.

4. The detector you built can also be used to determine whether solids conduct electricity. If the probes are touched to a solid and the LED lights up, then the solid conducts electricity. Electricity is not carried through solids by ions, as in solution, but by freely moving electrons in some types of solids such as metals. With your instructor's permission, you may use your detector to test if some solid substances conduct electricity.

Analyzing Evidence

1. What was the effect of varying the length of wire immersed in the solution? Explain any differences in the conductivity measurements.

2. What was the effect of varying the distance between the wires? Describe what could have contributed to any differences in the conductivity measurements.

3. What can you conclude about the presence of ions in each of the tested solutions?

4. Did the brightness of the light vary from solution to solution? Cite specific examples. What does a dim light tell us about the number of ions in a solution?

Interpreting Evidence

1. How do your results compare with your predictions? For cases where the measured conductivity is not what you expected, how have your investigation results caused you to revise your ideas about the structures of the compounds in these solutions?

2. How would a chemist describe what is present in salt water or hydrochloric acid solutions? What investigation evidence do you have that would support this description?

Making Claims

What can you claim about the presence of ions in various solutions?

Reflecting on the Investigation

1. If you were able to test some commercial products, read the labels on the containers of these liquids and attempt to determine what substances were responsible for the liquid's conductivity. Do you see any common types of ingredients in the liquids that were conductive?

2. The purity of ultrapure water is often measured with conductivity detectors. Based on what you have observed, why is this a good test for water purity? In what way might it be incomplete?

3. If you wanted to make a solution of an unknown solid substance to test it for the presence of ions, would it matter whether you used pure water or tap water to make the solution? Explain briefly.

4. What are some advantages and disadvantages of building your own lab equipment?

Analysis of Vinegar

Asking Questions

- How much acid is in vinegar?
- What is the acidic compound in vinegar?
- How can you measure acidity?

Preparing to Investigate

As described in Ch. 6 of *Chemistry in Context*, solutions of **acids** in water contain excess H^+ ions, while solutions of **bases** contain excess OH^- ions. Vinegar is a dilute solution of acetic acid, $HC_2H_3O_2$, in water.

A characteristic property of acids and bases is that they react with each other. Thus, acetic acid in vinegar will react with sodium hydroxide, a base, to produce water and sodium acetate.

acetic acid	+	sodium hydroxide	$\rightarrow$	water	+	sodium acetate
$HC_2H_3O_2$	+	NaOH	$\rightarrow$	H_2O+		$NaC_2H_3O_2$
(an acid)		(a base)				

This type of reaction is the basis for the **titration method of analysis** for acids. This method is described in detail in the introductory *Laboratory Methods* section of this lab manual. When using titration to analyze acids, a known quantity of acid solution is measured. Then, a solution of base is added slowly until just enough has been added to react with all of the acid. If the concentration of base is known and the volume of added base is carefully measured, you can calculate how much acid must have been present. Finally, if the volume of the acid solution is known, you can calculate the concentration of the acid. Concentrations are expressed in **molarity** (M), which is defined as the number of moles of a substance in exactly one liter of solution.

In a titration, an **indicator** is added so that a color change occurs to show when a reaction has taken place. The indicator used in this investigation is *phenolphthalein*, which is colorless in acid and pink in basic solutions. The end point of your titration will occur when one drop of base solution changes the solution from colorless to pink. Indicators are intensely colored so only one drop needs to be added to your titration investigation.

Making Predictions

- Look up and draw the structure of acetic acid, showing how the atoms are bonded together. What functional group does it contain? See Ch. 10 in *Chemistry in Context* for help. Speculate on why acetic acid is acidic. Which hydrogen makes the H^+? How does it react with NaOH?

- After reading *Gathering Evidence*, prepare a data table to record your titration results.

Gathering Evidence

Overview of the Investigation

1. Obtain a 24-well wellplate, four plastic transfer pipets, and necessary solutions.
2. Practice dispensing drops evenly.
3. Practice titrating vinegar with sodium hydroxide solution.
4. Titrate samples of vinegar to determine their acetic acid content.
5. Calculate the molarity and percent acidity of the acetic acid in the vinegar.

Part I. Setting up the Titration

1. Obtain your materials and label three pipets as follows: vinegar, NaOH, indicator. Fill the vinegar and NaOH pipets with the appropriate solution, and partially fill the "indicator" pipet with indicator.

    CAUTION! Sodium hydroxide solution is corrosive and can cause serious skin and eye damage. Therefore you MUST wear eye protection at all times and avoid getting any of the solution on your skin. In case of skin contact, rinse immediately with plenty of water and notify your instructor.

2. Follow the instructions given in the *Laboratory Methods* section to practice using a pipet and conduct a sample titration. It will help to place your wellplate on a white piece of paper in order to make the color change easier to see.

Part II. Measurement of the Acetic Acid Content of Vinegar

Remember that in doing titrations, your goal is to catch the point where *one drop* of NaOH produces the first permanent color change. Record your results on your data table.

1. Carefully add 20 drops of vinegar (count them) to each of six wells. Make sure that the drops fall directly to the bottom of the well and are not "trapped" along the side. (Why?) If you make a mistake, simply start again in another well. Record the position numbers for the wells you are using.

2. Add 1 drop of indicator to each of the six wells containing 20 drops of vinegar.

3. Start adding NaOH, one drop at a time, to the first well as you stir the solution with a toothpick. As you get close to the stopping point, you will find that the momentary pink color persists longer after each drop is added. Stop when one drop of NaOH produces a permanent pink or red color. Record the number of drops used. If you make a mistake, you may skip a trial and go on to the next one.

   NOTE: If you miss the end point, or lose count of drops, or are uncertain of the result for any other reason, it is good scientific practice **not** to erase the data. Simple make a note in your data table about what you think went wrong, draw a line through that row, and do another titration. All scientists make mistakes when doing lab work! Results crossed out will not affect your grade.

4. Repeat the titration for each of the other five wells, and record the number of drops used for each well.

   HINT: Once you know approximately how many drops of NaOH are needed, you can quickly add about 2/3 of that amount and add the remaining drops slowly until the end point is reached.

5. **Optional extension:** Analyze another kind of vinegar. First, rinse the vinegar pipet with the new sample twice by filling the pipet and emptying it into the appropriate waste container. Use a new set of wells in the wellplate for the new sample.

Clean-up

Your instructor will specify where to dispose of the contents of the wellplates and whether to wash or dispose of the plastic pipets. Wash the wellplate thoroughly and allow it to dry.

Analyzing Evidence

Recall that you will be using the known concentration of NaOH and the volumes of NaOH and vinegar that you used to calculate the concentration of acetic acid in your vinegar solution. Equation 1 shows the balanced equation for the reaction of acetic acid with sodium hydroxide:

$$HC_2H_3O_2 \ + \ NaOH \ \rightarrow \ H_2O \ + \ NaC_2H_3O_2 \hspace{3cm} (1)$$

This equation shows that *1 mole of acetic acid reacts with 1 mole of sodium hydroxide*. Thus, for the titration of vinegar, the number of moles of NaOH added from the pipet is exactly equal to the number of moles of acetic acid in the vinegar in the well. Expressed mathematically, we can say that

$$\text{moles of NaOH added from the pipet} \ = \ \text{moles of } HC_2H_3O_2 \text{ in the well} \hspace{1cm} (2)$$

The number of moles of NaOH in any sample of NaOH solution can be calculated from the molarity of the NaOH (expressed as moles per liter, or mol/L) and the volume (in liters), using the following equation:

$$\text{moles of NaOH used} = \text{liters of NaOH solution used} \times \frac{\text{moles of NaOH}}{1 \text{ liter of NaOH solution}} \hspace{1cm} (3)$$

A similar equation can be written for the moles of acetic acid in the vinegar sample.

$$\text{moles of acetic acid used} = \text{liters of vinegar} \times \frac{\text{moles of acetic acid}}{1 \text{ liter of vinegar}} \hspace{1cm} (4)$$

By substituting Equations 3 and 4 into Equation 2, we can write Equation 5:

$$\text{liters of NaOH} \times \frac{\text{moles of NaOH}}{\text{liters of NaOH}} = \text{liters of vinegar} \times \frac{\text{moles of acetic acid}}{\text{liters of vinegar}} \hspace{1cm} (5)$$

Or, writing this more succinctly,

$$\text{liters of NaOH} \times \text{molarity of NaOH} = \text{liters of vinegar} \times \text{molarity of acetic acid} \qquad \textbf{(6)}$$

This equation can be rearranged into a more useful form:

$$\text{molarity of acetic acid} = \text{molarity of NaOH} \times \frac{\text{liters of NaOH}}{\text{liters of vinegar}} \qquad \textbf{(7)}$$

There is one small problem that prohibits you from using Equation 6 directly – you don't know the volumes of your solutions in liters. You only know the number of drops of each solution that you used. If we assume that all of the drops from both pipets have the same volume, then the ratio of volumes expressed as drops should be the same as the ratio of volumes expressed as liters.

$$\frac{\text{liters of NaOH}}{\text{liters of vinegar}} = \frac{\text{drops of NaOH}}{\text{drops of vinegar}} \qquad \textbf{(8)}$$

Substituting Equation 8 into Equation 7 finally gives us Equation 9 for calculating the molarity of acetic acid.

$$\text{molarity of acetic acid} = \text{molarity of NaOH} \times \frac{\text{drops of NaOH}}{\text{drops of vinegar}} \qquad \textbf{(9)}$$

If you look at the label on the vinegar you used, you may see the amount of acetic acid reported not as molarity, but as *percent acidity*, which is defined as the number of grams of acetic acid per 100 mL of vinegar. One mole of acetic acid weighs 60 grams, and 100 mL is 1/10 of a liter; therefore the percent acidity can be calculated using Equation 10.

$$\text{percent acidity} = \frac{\text{moles of acetic acid}}{\text{liters of vinegar}} \times \frac{60 \text{ grams of acetic acid}}{1 \text{ mole of acetic acid}} \times {}^{1}\!/_{10} \qquad \textbf{(10)}$$

Calculations

1. Look carefully at the titration results recorded in your data table. Does any one result seem to differ greatly from the others? If so, a mistake may have been made, and it is legitimate to omit that result. Simply write a comment on your table next to that result saying that you are leaving it out.

2. Use all of the results that you think are valid to calculate the average number of drops of NaOH and record this on the data sheet.

3. Next, calculate the molarity of acetic acid in the vinegar, using the average number of drops of NaOH you used and Equation 9.

4. Lastly, calculate the percent acidity of your vinegar sample using Equation 10.

Interpreting Evidence

1. How does your calculated molarity compare to other acidic substances? Is it more or less acidic than stomach acid (about 0.16 M HCl)? How does it compare to acid rain? (see Ch. 6)

2. Why do you suppose that the acetic acid content in vinegar is reported as percent acidity rather than as molarity?

Making Claims

What can you claim about the acidity of vinegar based on your results from this investigation?

Reflecting on the Investigation

1. What is the advantage or doing several trials, especially since you are doing the exact same thing each time?

2. Suppose that during the titration, the addition of NaOH changed the color of the solution to a dark pink instead of a light pink. How would this change the results? Would your calculated molarity be higher or lower than the actual molarity? Explain.

3. Could you have done the titration "backwards," starting with 20 drops of NaOH, adding the phenolphthalein indicator, and then adding the vinegar one drop at a time? Explain. What would you observe? Would the endpoint be easier or harder to see than with the titration method you used?

4. Can you think of ways to improve this method of analysis? What changes might be possible to give you more accurate results?

5. Titration can be used to determine the acid content of any food or drink. List three that you believe contain acid. (Hint: Acidic foods often taste tart.) Why would titrations of highly colored substances require some modification to the procedure?

Notes

Measuring Water Hardness

Asking Questions

- What ions are present in water and in what concentrations? Which of these make the water "hard"?
- What problems can result from having hard water?
- How do water softeners work?

Preparing to Investigate

In this investigation, you will analyze water samples to determine the combined concentrations of calcium and magnesium ions, which is referred to as the "total hardness" of the water. You will use a titration method of analysis, which is described in detail in the introductory *Laboratory Methods* section of this text.

Your analysis will be based on the chemical reaction of Ca^{2+} and Mg^{2+} with an ion called dihydrogen ethylenediaminetetraacetate. This ion can be abbreviated EDTA and has the molecular formula $C_{10}H_{12}N_2O_8^{4-}$. In the chemical reaction you will perform, the EDTA ion has two protons (H^+) attached, which reduces the charge from 4- to 2-. The ion with the extra protons is often written H_2EDTA^{2-}. The reaction you will do involves one calcium or magnesium ion reacting with one H_2EDTA^{2-} ion, as shown below.

$$Ca^{2+} + H_2EDTA^{2-} \rightarrow Ca(EDTA)^{2-} + 2 H^+$$

To do the titration, you will measure out a known amount of your water sample, and then titrate the sample with EDTA solution until it completely reacts with the calcium and magnesium ions in the sample. In order to see the reaction, you will use a small amount of an **indicator** called calmagite. At pH 10 (an alkaline solution), the indicator has a deep blue color, but in the presence of metal ions it changes to red. Thus, if a drop of indicator is added to a solution containing calcium and magnesium ions, the solution will have a reddish color. When enough EDTA has been added to react with all of the calcium and magnesium, the solution will turn purple and then blue.

Your strategy for this investigation will be to first use an EDTA solution to titrate a reference calcium solution of known concentration, and then use the same EDTA solution to titrate an unknown water sample. Comparing the two results will allow easy calculation of the hardness of the unknown sample. For simplicity, the amounts of Ca^{2+} and Mg^{2+} that are present in the water sample will be measured together and represented by a single concentration.

The common practice in water hardness is to report the results as *milligrams of CaCO₃ per liter of water.* This does not mean that solid calcium carbonate is actually present, but rather that this amount of solid calcium (or magnesium) carbonate could be formed from the amount of Ca^{2+} (and Mg^{2+}) present in the water.

Making Predictions

- Look at a periodic table and identify the locations of magnesium and calcium. From the discussion of atomic structure in Chapter 1 of *Chemistry in Context*, explain why these two elements may react in a similar manner to each other in this investigation.
- After reading *Gathering Evidence*, prepare a data table to record your titration results.

Gathering Evidence

Overview of the Investigation

1. Test the indicator color change.
2. Do three or four titrations of water containing a *known* concentration of calcium ions.
3. Calculate the calibration factor for the EDTA solution.
4. Do three titrations for each of several water samples containing an *unknown* concentration of calcium and magnesium ions.
5. Calculate the water hardness for the unknown samples.

 STOP! Safety glasses must be worn *at all times* while doing chemistry investigations.

Part I. Setting up the Titration

1. Obtain two small, clean and dry test tubes and a rack to hold them. Label them and fill them about 1/3 full with each of these two solutions:

    a. EDTA solution

    b. Reference solution of known hardness – this will be approximately equivalent to 0.500 mg of $CaCO_3$ per mL of water. Be sure to note the exact concentration provided by your instructor.

2. Label and fill four clean and dry plastic pipets as follows. *It is important to label the pipets so that you do not get them mixed up!*

    a. Label one as "buffer" and fill it 1/3 full with a pH 10 buffer solution.

    b. Label another as "indicator" and fill it 1/3 full with the calmagite indicator.

    c. Label a third pipet as "EDTA" and place it in the EDTA test tube.

    d. Label the last pipet as "Ca reference" and place it in the test tube with the reference solution.

3. Place a 24-well plastic wellplate on a piece of white paper to aid in seeing the color changes. As you are doing the investigation, be sure to note the letter and number of each

well on the plate that you use, and record which titration investigation is conducted in each well.

Part II. Titrations of the Reference Solution

1. If necessary, practice using the pipet. Make sure that you can dispense complete drops, as outlined in the *Laboratory Methods* section, and also that you can fill a 1-mL graduated-stem pipet *exactly* to the 1-mL line. The best method for doing this is to squeeze the air out of the bulb, insert the pipet tip into the solution, and gradually release the pressure, observing how high the solution rises. Remove the pipet tip from the solution when the liquid is at the 1-mL line, taking care not to change the pressure on the bulb.

2. Test the color change of the indicator so that you know what to expect during your titration. In one well of the wellplate, add 10 drops of pure water, and then 2 drops of pH 10 buffer and 1 drop of calmagite indicator. Stir the mixture with a small stirring rod and note the color. If it is not blue, add 1 drop of EDTA solution to produce a bright clear blue color with no trace of red or purple. This is the color you will be looking for at the end of each titration. If you cannot get this color then check with your instructor before proceeding further. Leave this solution in your wellplate as a reference for when you are doing your titrations.

3. Dispense exactly 1 mL of the calcium reference solution into each of four wells in the wellplate. Into the first well, add 2 drops of buffer and then 1 drop of indicator, in that order. Then, add the EDTA solution slowly, one drop at a time, counting drops and stirring after each addition. Continue until the color starts to change from red to purple. Wait a few moments to see if the color continues to change (the reaction is slow); then *slowly* add additional drops, stirring after each one, until the color becomes pure blue. Keep track of the total number of drops added, and when you reach the endpoint, enter the number of drops in your data table.

4. Repeat the titration procedure for the other three wells containing the calcium reference solution, titrating one sample at a time to the endpoint. Remember to add 2 drops of buffer and 1 drop of indicator (always in that order) before each titration.

5. Do additional titrations if you are uncertain of the results of any of your titrations (for example, if you lost any drops of solution, think that you went past the end point, or lost count of the drops you used). With a little practice, the titrations can be done quite rapidly. (The slowest part of this particular titration is waiting for the chemicals to react completely.)

Part III. Titrations of Unknown Water Samples

Obtain one or more water samples to be analyzed, as directed by your instructor. These might be the tap water in your laboratory, dormitory or home. They might be samples from a nearby lake or river, or from a well or spring.

NOTE: Some water samples – especially tap water in some buildings – may not change color to pure blue. This is caused by traces of iron or copper in the water. If this is the case with your sample, check with the instructor before proceeding further. It may be necessary to titrate to a purple color rather than a pure blue.

1. Label a clean 1-mL graduated-stem pipet for your water sample.

2. Carefully dispense 1 mL of your water sample into a clean well in the wellplate, and note which well you use on your data table.

3. Add two drops of buffer followed by 1 drop of calmagite indicator, and stir. Titrate your water sample exactly as you did for your reference solution. Since you don't know how many drops will be required, you should do this titration slowly.

4. Do two or three additional titrations on this sample, as indicated by your instructor. Depending on how much EDTA you used in your first trial, you may want to use a larger or smaller volume of the unknown sample for the remaining titrations. If you use a different volume, be sure to record the volume you use in your data table. Remember to add buffer and indicator each time, and record the number of drops for each trial in your data table.

Part IV. Optional Extensions

You may want to analyze additional water samples, or your instructor may assign another sample to be analyzed. If so, record the sample information on your data table. Use a new pipet or rinse your sample pipet thoroughly before using it with a new sample. Below are some ideas for further investigation.

1. Take 10-20 mL of water that is relatively "hard" and run it slowly through a small tube containing ion-exchange resin beads. Collect the water in a test tube and titrate it using the procedure outlined above. Is the hardness of the water different? Is this what you expected?

2. Your instructor may set up a distillation apparatus in the laboratory. Chapter 5 in *Chemistry in Context* shows a similar apparatus. Tap water or other hard water will be placed in the heated flask and distilled. Place a small sample of the distilled water in a clean test tube and titrate it for hardness. How does the hardness of distilled water compare with the starting material?

3. Your instructor may ask you to analyze some water samples for calcium and magnesium using modern analytical instrumentation. You will be given detailed instructions on how to use the instrument and do the accompanying calculations. You can then compare the instrumental results with those obtained by the titration procedure.

Clean-up

Dispose of solutions and pipets in appropriate containers as instructed. Wash your wellplate and allow it to drain.

Analyzing Evidence

The reference solution should have a "hardness" close to 500 mg/L, or 0.500 mg/mL. Using the first set of titrations, you can calculate the *calibration factor*, or the hardness corresponding to 1 drop of EDTA solution, using Equation 1.

$$\frac{\text{mg hardness of reference solution}}{1\ \text{mL reference solution}} \times \frac{1\ \text{mL reference solution}}{\text{average drops of EDTA}} = \frac{\text{mg of hardness}}{1\ \text{drop of EDTA}} \quad \textbf{(1)}$$

For the analysis of an unknown water sample, multiply the average number of drops of EDTA used by the calibration factor to obtain the mg of hardness per 1 mL of water. Equation 2 assumes that you used 1 mL of your sample; if you used a different volume you should put that volume into the equation.

$$\frac{\text{drops of EDTA}}{1\ \text{mL of sample}} \times \frac{\text{mg of hardness}}{1\ \text{drop of EDTA}} = \frac{\text{mg of hardness}}{1\ \text{mL of sample}} \quad \textbf{(2)}$$

Finally, covert your calculated result into the standard units of *mg of hardness per liter of water*. You can do this using the conversion factor 1 L = 1000 mL. You can then rate the hardness of each water sample analyzed, using the following classification.

Description	Concentration
Very hard	Over 300 mg/L of $CaCO_3$
Hard	150 to 300 mg/L of $CaCO_3$
Moderately hard	50 to 150 mg/L of $CaCO_3$
Soft	0 to 50 mg/L of $CaCO_3$

Share your results with the other students in your class. Your instructor will provide details about how to share your data. If everyone has analyzed the same sample, the results can be assembled and compared. If the class was divided up to analyze various samples, you should record what other students found for their samples and compare the results to yours.

Calculations

1. Look carefully at the data from each of the sets of titrations to see whether the results show consistency or whether any one result in a given set should be eliminated because it appears to be an outlier. If so, draw a single line through the result in your data table and make a note that you have not included it in your analysis.

2. Calculate the average number of drops of EDTA used for each set of titrations.

3. Using your results from your first set of titrations and Equation 1, calculate the calibration factor for your investigation.

4. Calculate the hardness of each water sample you titrated, using Equation 2. Convert the units to *mg of hardness per liter of water*, and classify each sample according to the table.

Interpreting Evidence

1. Based on your own data and/or class data, how would you classify the hardness of your local tap water?

2. If your class analyzed other water samples (lake, river, well water, etc.), did you find significant differences in hardness? If so, suggest possible explanations.

3. If you investigated the use of an ion exchange resin, how effectively did it soften the water? What were the levels of hardness before and after treatment?

4. If you analyzed a sample of water from a distillation apparatus in the lab, what can you conclude about its effectiveness in removing calcium and magnesium ions from the water?

Making Claims

What can you claim about the differences in water hardness from different geographical areas, water sources, or water treatment processes?

Reflecting on the Investigation

1. Write a summary paragraph describing the causes of hard water, the symptoms you might notice if your household has hard water, and ways to soften water.

2. Is very soft water the same as very pure water? Explain.

3. It is important in scientific investigations to quantify the errors that may be present in your measurements. In these titrations, there may be an uncertainty of at least one drop in identifying the exact end point. If you use 20 drops of EDTA solution in a titration, what percent uncertainty in the hardness is contributed by adding one extra drop?

$$\% \text{ error} = \frac{\text{number of drops you are in error}}{\text{number of drops used}} \times 100\%$$

Measuring Chloride in Water Samples

Asking Questions

- How does human activity affect water purity?
- Why does chloride ion concentration in water increase with human activity and pollution?
- How does chloride ion differ from elemental chlorine?

Preparing to Investigate

Water contains many dissolved ions, including chloride (Cl^-), an anion found in water and sewage. The chloride content of natural, unpolluted surface waters depends in large part on the geology of the area. In places where surface water normally has very little chloride, a higher chloride concentration implies a source from human activity. In this investigation, you will measure the chloride concentration in water samples taken from streams or rivers in your local area and also, if possible, samples from your local wastewater treatment plant. By dividing up the class to analyze different samples, it will be possible to collect a large quantity of data in a short time. You will then use the results to draw conclusions about the human impact on local surface waters.

Note that chloride (Cl^-), an ion, is quite different than elemental chlorine (Cl_2), a gas. Chlorine is added as a disinfectant to drinking water in small amounts (less than 1 ppm) and to swimming pools in somewhat larger amounts (about 5 ppm). Chlorine will produce some chloride as it reacts, but in most cases it is not a major source of chloride.

You will measure the chloride content in water by performing a **titration** in which the chloride reacts with silver nitrate ($AgNO_3$) to form an insoluble white compound, silver chloride ($AgCl$). Other ions that are present in the water do not participate in this reaction.

$$AgNO_3(aq) + Cl^- \text{ (in the water sample)} \rightarrow AgCl(s) + NO_3^-(aq)$$

In order to know when enough silver nitrate has been added to react with all of the chloride in your water sample, you will use a solution of sodium chromate (Na_2CrO_4) as an **indicator**. Chromate ions are yellow, but they react with silver ions to form a red precipitate of silver chromate.

$$2\ Ag^+(aq) + CrO_4^{2-}(aq) \rightarrow Ag_2CrO_4(s)$$

Thus, as silver nitrate is added to your water sample, the chloride is precipitated as white silver chloride. After all of the chloride has been removed, additional silver ion will react with chromate to form a red insoluble precipitate of silver chromate. The appearance of this red precipitate signals the endpoint of the titration. If you have not yet performed a titration in this laboratory course, you should review the section about titration in the *Laboratory Methods* chapter.

Making Predictions

- After reading *Gathering Evidence*, prepare a data table to record your results. Do you expect your tap water sample to have a high or low concentration of chloride? Do you expect your assigned or collected water sample to have a higher or lower concentration of chloride than tap water?
- Sodium chromate, the indicator used in this investigation, is a carcinogen. Define the term *carcinogen*. What precautions should you take when working with this compound?

Gathering Evidence

Overview of the Investigation

1. Collect water samples at various locations
2. Practice doing chloride titrations with tap water and silver nitrate.
3. Calculate the concentration of chloride in the tap water sample.
4. Titrate the collected water samples with silver nitrate.
5. Calculate the concentration of chloride in the collected samples.
6. Compare the class data for different water samples.
7. Draw conclusions about human impact on the local aquatic environment.

Part I. Collecting Water Samples

An important aspect of the investigation is choosing sample locations and collecting water samples from those locations. Before the lab class, your instructor will assign a sample location for each pair of students (or will arrange to have samples collected and made available in the laboratory). If a major river or stream flows through your town or city, then water samples should be collected upstream of the city, in the middle of the city, and downstream of the city. Make sure that the downstream sample is collected far enough away from the city that any sewage treatment plant effluent has mixed freely into the river. Samples should also be collected from the sewage treatment plant itself. Both the effluent and the influent (which should be aerated because of its high suspended solid content) should be sampled. Your class also should obtain samples from the municipal drinking water sources, such as reservoirs or lakes. Finally, it is helpful, for comparison purposes, to collect samples from sources, such as ponds or small streams, that are "clean" in that they are expected not to have been impacted by human activity., If you collect your own sample(s), be sure to record accurately the location of the source and label the collection vessel(s).

Part II. Setting up the Titration

1. Obtain a plastic wellplate with 24 wells and a small stirring rod. Place the wellplate on a piece of white paper so that the color changes can be seen easily.
2. Obtain two plastic pipets. Label one "water sample" and the other "silver nitrate" or "AgNO₃". *Labeling is crucial because it is easy to get them mixed up!*

3. Dispense about 10 mL of silver nitrate solution into a clean *dry* beaker or other container. Label the container and record the exact concentration of the solution in your data table.

 CAUTION! Silver nitrate will stain your skin black, so be careful not to get it on your hands. The stain is harmless and will wear off in a few days, but in the meantime, its appearance is unattractive.

4. Obtain your water samples for analysis and a beaker of tap water.

5. Perform a practice titration using the tap water:

    a. Use your "water sample" pipet to add 10 drops of tap water to a well in the wellplate.

    b. Add 1 drop of sodium chromate indicator and stir. The mixture should be pale yellow-green.

 CAUTION! Sodium chromate is a carcinogen. Avoid contact with your skin.

    c. Use your other pipet to add $AgNO_3$ solution, *one drop at a time*, with gentle stirring. Count the drops and carefully observe what happens. As each drop is added, you will observe a cloudy appearance as solid silver chloride is formed. A reddish color may appear and then disappear on stirring. As you near the endpoint, the mixture in the well will have an orange tint, and the first appearance of a uniform tint that does not disappear with stirring signals the end point. Make a note of how many drops were used. Then, add 1 or two more drops of silver nitrate with stirring. The color will become darker orange or red due to formation of more Ag_2CrO_4.

    d. Finally, add a few more drops of tap water, with stirring, until the color reverts back to a milky yellow-green. Save this mixture for color comparison.

Part III. Analyzing Tap Water

In doing titrations, your goal is to catch the point where one drop of $AgNO_3$ produces the first permanent color change. It is not a dramatic color change but rather a slight orange tint. This is the *endpoint* where all of the chloride has been used up by reaction with the $AgNO_3$. If you are uncertain of your ability to use the pipet to dispense a set number of same-sized drops, you should follow the procedure for pipet practice in the *Laboratory Methods* section.

1. Carefully add 20 drops of tap water (count them) to each of four wells in your wellplate. Record the position numbers for the wells you are using on your data table.

2. Add 1 drop of indicator to each well.

3. Start adding $AgNO_3$, one drop at a time, to the first well as you stir the solution. Stop when one drop of $AgNO_3$ produces a permanent orange color. Record the number of drops you used on the data sheet. If you make a mistake, skip this trial and go on to the next one.

HINT: The gradual color change from yellow to pale orange may be hard to recognize. If so, it is useful to use your practice titration mixture for comparison. You are looking for the first change away from that color.

4. Repeat the titration for each of the other three wells, and record the number of drops you used for each well in your data table. With a little practice, the titrations go very quickly.

HINT: Once you know approximately how many drops of $AgNO_3$ are needed, you can quickly add about 2/3 of that amount, and then slowly add the remaining drops to the endpoint.

NOTE: If you missed the endpoint or lost count of drops, or are otherwise uncertain of a result for any reason, it is good scientific practice **not** to erase the data. Make a note on your data table about what you think went wrong, draw a line through that column, and do another titration. Crossed-out results will not affect your grade for the investigation.

5. If you have chosen to omit any of your trials or your data seem scattered, then you may wish to titrate an additional one or two samples of tap water.

Part IV. Analyzing Collected Water Samples

1. Rinse the pipet and beaker used for tap water, first with distilled or deionized water, and then with the new sample to be analyzed.

2. Do a trial titration with 20 drops of the new sample. Depending on the nature of the sample, it may have very little chloride, requiring only a few drops of silver nitrate; or the chloride concentration may be very high, requiring a large quantity of silver nitrate. Record the number of drops of silver nitrate used.

3. Depending on the results from this first titration, you may decide to use more than 20 drops or less than 20 drops of your water sample. Dispense samples of the desired size into three more wells. Record all of your data in your data table.

4. If time permits, you can analyze additional water samples.

Clean-up

Dump the contents of the wellplate into a designated waste container. Rinse the wellplate with tap water, then with distilled or deionized water if available. Leave it upside down on a paper towel to drain. Your instructor will indicate whether the pipets are to be discarded or saved.

Analyzing Evidence

After performing the titrations, you should have at least four results for tap water and four results for at least one other water sample. The same calculation procedure is used for both sets of data.

The balanced chemical equation for the reaction of chloride ion with silver nitrate shows that 1 mole of Cl^- reacts with 1 mole of $AgNO_3$.

$$AgNO_3(aq) + Cl^- \text{ (in the water sample)} \rightarrow AgCl(s) + NO_3^-(aq) \qquad \textbf{(1)}$$

Thus, for the titration of a water sample, the number of moles of silver nitrate added equals the number of moles of chloride ions in the water sample that was in the well.

$$\text{moles of chloride in the well} = \text{moles of } AgNO_3 \text{ added} \qquad \textbf{(2)}$$

The number moles of a substance in a solution equals the volume of that solution multiplied by the concentration of the solution in moles per liter (molarity). Thus,

$$\text{moles of } AgNO_3 \text{ added} = \text{molarity of } AgNO_3 \times \text{volume of } AgNO_3 \qquad \textbf{(3)}$$

and

$$\text{moles of chloride in water sample} = \text{molarity of chloride} \times \text{volume of water sample} \qquad \textbf{(4)}$$

Substituting Equations 3 and 4 into Equation 2 gives Equation 5.

$$\text{molarity of chloride} \times \text{volume of water sample} = \text{molarity of } AgNO_3 \times \text{volume of } AgNO_3 \quad \textbf{(5)}$$

If we assume that the drops added are all the same volume, we can say that

$$\text{molarity of chloride} \times \text{drops of water} = \text{molarity of } AgNO_3 \times \text{drops of } AgNO_3 \qquad \textbf{(6)}$$

Rearranging gives Equation 7, the final working equation.

$$\text{molarity of chloride} = \text{molarity of } AgNO_3 \times \frac{\text{drops of } AgNO_3}{\text{drops of water sample}} \qquad \textbf{(7)}$$

Equation 7 will give you the concentration of chloride in molarity, or mol/L. Convert the concentration to milligrams of chloride per liter of water (mg/L), which is the same as parts per million (ppm). To do this, multiply the molarity of the chloride by the molar mass of chloride ion – 35.5 g/mol, or 35,500 mg/mol.

Calculations

1. Look carefully at the four titration volumes for tap water. Does any one result seem way out of line from the others? If so, it is likely that a mistake was made, and it is legitimate to omit that result. Simply cross out the result and write a note about what you think went wrong. Use all of the results that you think are valid to calculate the average number of drops of $AgNO_3$ used to titrate your tap water.

2. Use Equation 7 above to calculate the molarity of chloride in the tap water.

3. Finally, convert the answer to milligrams of chloride per liter of water (mg/L) or parts per million (ppm).

4. Repeat the calculations for the other water samples you analyzed, and share your results with your classmates as specified by your instructor.

Interpreting Evidence

1. What patterns do you see in the assembled data from your class?

2. After examining the assembled data, what can you conclude about the impact of your city on the local river or stream? Does human activity seem to add chloride to the water?

3. What is the advantage of doing three or more trials, especially since you are doing exactly the same thing each time?

4. Suppose that during the titrations you consistently added sufficient $AgNO_3$ to change the color of the indicator to a darker orange-red. Would this change the results? Would your measured chloride concentrations be higher or lower than the actual concentrations? Explain.

Making Claims

What can you claim about how human activity affects water quality?

Reflecting on the Investigation

1. List some of the major non-industrial sources of chloride in natural waters. Which ones do you think are important in your area?

2. Based on what you have learned in this investigation, do you consider chloride a serious water pollutant? Why or why not?

Analyzing Bottled Water

Asking Questions

- What compounds, other than water, are found in bottled water?
- What are some reasons that people drink bottled water?
- What are some problems with the consumption of bottled water?

Preparing to Investigate

Bottled water is rarely, if ever, pure water. As discussed in Chapter 5 of *Chemistry in Context,* it often comes from wells or springs that give it a high mineral content, particularly calcium and magnesium ions. If you performed Investigation 14, you know that the presence of these ions contributes to the **total hardness of water**, and their concentrations can be measured using titration analysis. Calcium and magnesium are often accompanied in the water by bicarbonate ion, HCO_3. Bicarbonate is a base, and it contributes to the **alkalinity** of the water. Its concentration can also be measured using a titration investigation. In this investigation, you will determine the hardness and alkalinity of water. You also will measure the chloride concentration of your sample (as in Investigation 15).

The pH of water must be close to neutral, pH 7, in order for the water to be safe for human consumption. However, if the water contains a substantial concentration of bicarbonate it will be more alkaline, with a pH above 7. You will measure the pH using a pH meter.

Another way to characterize the water sample is to determine the total amount of dissolved material, often referred to as **total dissolved solids (TDS)**. TDS consists mostly of ionic substances, or salts, since water is an excellent solvent for ions. Some bottled-water labels indicate the TDS concentration. You will measure TDS by heating the sample to evaporate the water and determining the mass of any solid material that remains behind.

For easy comparison, all investigation results except for pH will be converted to parts per million (ppm, or milligrams of a substance per liter of water). Because of the large number of analyses required in this investigation, you will be assigned one sample and some analyses to perform, and data from your class will be assembled for comparison and discussion. You should base your claims and reflections on the data from the whole class.

Making Predictions

After reading *Gathering Evidence*, make a table to record the data you will collect in this investigation. How do you think the hardness of the bottled water will compare to the samples you measured in Investigation 16? How do you think the chloride concentration of the bottled water will compare to the samples you measured in Investigation 17?

Gathering Evidence

Overview of the Investigation

1. Measure the pH of the water using a pH meter.
2. Measure the concentration of calcium and magnesium by titration.
3. Measure the bicarbonate ion concentration by titration.
4. Measure the chloride ion concentration by titration.
5. Measure the total dissolved solids in the water.
6. Assemble the analytical information to compare results for different samples.

Part I. Measuring the pH

This should be done following the procedure described in the *Laboratory Methods* chapter of this book. If you have not yet used a pH meter, your instructor will demonstrate it. It is important to learn the proper use and operation of this instrument so that you do not damage it.

Alternatively, you can use pH test strips to measure the pH of your samples. This is fast, but less precise, than using a pH meter.

Record the pH of your sample on your data sheet.

Part II. Concentration of Calcium and Magnesium

Measure the concentration of calcium and magnesium, or the hardness of the water, using the titration procedure described in Investigation 16. You will titrate your sample with a solution of EDTA and use calmagite as an indicator to determine when the endpoint has been reached.

NOTE: If your instructor provides an EDTA solution of known concentration, there is no need to do the initial titrations of the reference solution as described in Investigation 16. In that case, a specific number of drops of your water sample (e.g. 10 drops) should be used rather than measuring exactly 1 mL of the sample. You may need to do several titrations using different numbers of drops of your water sample to determine the best amount to use for good results. If this is done, the calculation is done using the following equation.

$$\text{molarity of Ca} + \text{Mg} = \text{molarity of EDTA} \times \frac{\text{drops of EDTA solution}}{\text{drops of water sample}} \qquad \textbf{(1)}$$

For an explanation of the logic behind this calculation, refer to Investigation 15 or 17.

To convert your calculated molarity into ppm, you need to multiply the molarity by the molar mass of the ion. This analysis measures the *combined* concentration of calcium and magnesium, but since calcium is almost always present in much higher concentration than magnesium we will simplify the calculation by assuming that all of the measured hardness is due to calcium. Do the conversion by one of the following methods:

- If you calculated the molarity of Ca + Mg by equation 1 above, multiply this molarity by 40,000 mg/mole.

- If you used the calculation procedure from Investigation 16, multiply your calculated "hardness" by 0.40 to convert from mg $CaCO_3$ per L to mg Ca per L. The conversion factor comes from the ratio of the atomic weight of Ca to the molecular weight of $CaCO_3$.

Part III. Concentration of Chloride

Measure the concentration of chloride ion in the water using the procedure described in Investigation 17. You will titrate your sample with silver nitrate and use sodium chromate as an indicator to show when the endpoint has been reached. Calculate the concentration of chloride ion in ppm exactly as specified in Investigation 17.

Part IV. Concentration of Bicarbonate

Analysis of bicarbonate is not described elsewhere in this laboratory manual, so a more detailed introduction is given here. The procedure is a titration analysis, based on the fact that bicarbonate is a base that can react with an acid such as hydrochloric acid (HCl) as shown here.

$$HCO_3^- \text{ (aq)} + HCl \text{ (aq)} \rightarrow H_2O \text{ (l)} + CO_2 \text{ (g)} + Cl^- \text{ (aq)}$$

A colored indicator, methyl orange, will change color from yellow to orange when the pH drops below 4. This occurs when an excess of HCl is present and indicates that all of the bicarbonate ion has reacted. If this is your first time doing a titration or you need a refresher on how to do one, read the section on titrations in the *Laboratory Methods* section of this lab manual.

1. Gather the materials you need for your titration, including a plastic wellplate, small stirring rod, and two pipets. Place the wellplate on a white sheet of paper so that the color change can be seen easily, and label your pipets "water sample" and "HCl". Labeling is crucial because it is easy to get the pipets mixed up! If necessary, practice using the pipet so that you can reliably dispense one full drop of liquid at a time into the wellplate. A pipet practice procedure is given in *Laboratory Methods*.

2. Obtain about 10 mL of HCl solution in a clean, dry beaker and record its concentration.

3. Add 10 drops of your water sample into a well of the wellplate (record the well position), and then add 1 drop of methyl orange indicator. Your solution should be yellow.

4. Add HCl solution, *one drop at a time*, counting the drops and stirring gently after each addition. Carefully observe what happens. Watch for the *first color change* from yellow to orange that persists after the mixture is stirred, and record the number of drops used.

5. To aid in recognizing the color change for subsequent titrations, add a few more drops of bottled water to the well. The color should go back to yellow. Save this for color comparison. In your other titrations, you will be looking for the first color change away from this yellow.

6. From the result of the first titration, decide how much water to use for subsequent titrations so as to require a convenient amount of HCl. If possible, you should use at least 10 drops but no more than 30 drops of HCl.

7. Carefully add the desired number of drops of water to three more wells in your wellplate.

8. Add 1-2 drops of methyl orange indicator to each well.

9. Titrate each of the three samples, one at a time, using the HCl solution. Record the number of drops used. If the results do not show adequate consistency, do one or two additional titrations.

10. Average your results and calculate the amount of bicarbonate present in the sample using Equation 2. The equation is derived using the same logic as described for the chloride ion concentration in Investigation 17.

$$\text{molarity of HCO}_3^- = \text{molarity of HCl} \times \frac{\text{number of drops of HCl solution}}{\text{number of drops of water sample}} \qquad (2)$$

To convert molarity to ppm, multiply your calculated molarity by the molar mass of bicarbonate, 61,000 mg/mole.

Part V. Concentration of Total Dissolved Solids

1. Check that a large hot plate has been turned on and is ready to use.

2. Label two clean, dry 50-mL beakers with your sample number and your initials. Confirm that the balance reads 0.000 g with nothing on the pan. Weigh each beaker and record the mass to the nearest milligram (0.001 g). The use of an analytical balance is described in detail in the *Laboratory Methods* section.

3. Use a graduated cylinder to measure 20 mL of bottled water and add it to one of your weighed beakers. Repeat the process for the other beaker.

4. Place the beakers on the hot plate. Adjust the temperature so that the water will boil gently. Continue heating until the water has *completely* evaporated from the beakers.

5. Use tongs (*the beakers will be HOT!*) to remove the beakers from the hot plate and allow them to cool. Carefully inspect the contents to ensure that the beakers are completely dry and there is not any moisture remaining on the bottom or sides.

6. Reweigh the beakers and record the masses.

7. Average your results and calculate the TDS as follows. First, convert the volume of water used to liters (e.g. 20 mL is 0.020 L).

$$\text{TDS in ppm} = \frac{\text{g of solid recovered} \times 1000 \text{ mg/g}}{\text{volume of water (in L)}}$$

Clean-up

When finished, follow all instructions regarding waste disposal. Put solutions into appropriate waste containers. Your instructor will tell you whether to wash or dispose of the plastic pipets. The remaining glass and plastic items should be rinsed thoroughly with deionized or distilled water and left upside-down to drain.

Analyzing Evidence

1. With your lab partners, complete the necessary calculations for all five sections of the analysis. Remember that aside from pH, all other measurements should be in units of ppm.

2. Share your results with others from your class by writing them on the board or following another procedure specified by your instructor, and record your classmates' results.

Interpreting Evidence

1. The concentrations of two negatively charged ions, chloride and bicarbonate, were measured. Which was present in higher concentration in the bottled water sample? To answer this question you should compare *molarities*, not ppm.

2. Using your results, what is the ratio of the bicarbonate *molarity* to the combined *molarity* of calcium + magnesium? If the calcium + magnesium and bicarbonate all came from limestone dissolved in the water, then the bicarbonate molarity should be exactly twice that of the calcium + magnesium molarity. How closely do your results agree with this prediction?

3. The pH of water for human consumption should be close to neutral. How closely does your bottled water sample compare?

4. Compare the different water samples analyzed by your classmates to yours. Can you see any differences based on water treatment, source, or other characteristic of the water?

Making Claims

What can you claim about the compounds contained in bottled water?

Reflecting on the Investigation

1. Check the labels on the bottled water(s) that you studied. Were they treated or purified in some way? If so, list the methods used. What was the specific goal of the treatment? If necessary, see Ch. 5 in your textbook for help.

2. How does the hardness of your water compare to the samples you measured in Investigation 16?

3. How does the chloride concentration compare to the samples you measured in Investigation 17? Does the difference make sense based on what you know about the sources of chloride in water?

4. Do you feel that drinking bottled water is more, less, or equally safe than drinking tap water? Write a paragraph explaining your position.

Notes

Reactions of Acids with Common Substances

Asking Questions

- What is an acid?
- What sorts of elements or compounds react with acids?
- How and why does acid rain affect buildings and living organisms?

Preparing to Investigate

Acids are substances that contain hydrogen ions (H^+) or react with water to form hydrogen ions. An important category of chemical substances, acids are described in detail in Chapter 6 of *Chemistry in Context*. They are found in many places, including certain foods, our bodies, and frequently in rain or snow. Acids undergo chemical reactions with a great variety of substances. Some of these reactions are desirable, while others can be quite damaging. For instance, the acids found in rain or snow can have destructive effects on various building materials and other adverse environmental consequences.

In this investigation, you will observe the reactions of three common acids: hydrochloric acid, sulfuric acid, and nitric acid. While all of these compounds are acids, you will see that not all acids react in exactly the same way. Careful observation will also show that the rate of reaction of acids with materials varies with concentration. In order to make observations in a limited amount of time, you will use acid solutions that are far more concentrated than those found in rain or snow. Still, the same reactions do occur with acid rain but on a vastly slower time scale.

You will study the reactions of these acids with some familiar substances. You will observe reactions with four common metals – zinc, copper, nickel, and aluminum. Most metals react with acids to produce hydrogen (H_2), a colorless gas, and metal ions. You will also see the effects of acid on marble, a form of calcium carbonate. Building materials vary widely in their reactivity with acid; some, like marble, react readily with acids while others, such as granite, do not visibly react at all. Lastly, you will see the reaction of acid with egg white, a protein that represents living matter. The reaction of proteins with acid is unlike that of the other materials. In aqueous solutions, protein molecules have a preferred three-dimensional shape that is pH dependent. Addition of an acid lowers the pH and changes the protein shape so that the protein becomes less soluble and formation of a solid may be visible. In addition, some acids will chemically react with the protein molecule itself.

Concentrations of the acids in this investigation are expressed as **molarity.** The molarity of a substance in solution is defined as the number of moles of that substance in 1 liter of solution. It is expressed in units of mol/L or M.

Making Predictions

- Locate zinc, copper, nickel and aluminum on a periodic table. Do you expect their reactions with acid to be similar to each other or different? Why?
- Do you expect to see greater reaction from 6 M or 0.6 M sulfuric acid? Why?
- After reading *Gathering Evidence*, prepare a table to record your observations from the investigation of the four different acid solutions with the six different samples.

Gathering Evidence

Overview of the Investigation

1. Test the reactions of four acid solutions with marble chips (calcium carbonate).
2. Test the acid solutions with four different metals.
3. Test the acid solutions on egg whites.

 STOP! You should always wear eye protection in the laboratory, but this is especially important when working with acids. Safety glasses are absolutely essential to protect your eyes from any splashes.

Part I. Preparing the metal samples.

The tests with metals will work best when the metal surface has been freshly cleaned to remove any corrosion or coating. Therefore, you should obtain the following sets of metal pieces and clean at least one surface of each piece with sandpaper, emery paper, or steel wool.

- Zinc: These samples can be either small, thin strips of zinc or galvanized nails, which are made of iron coated with zinc. When clean, the surface should be bright and shiny.
- Copper: These samples may be very short pieces of heavy-gauge copper wire or small pellets of copper shot. The surface should be bright and shiny.
- Nickel: Fresh, shiny paper clips are usually made of iron or steel with a coating of nickel. If the paper clips are new and shiny, they shouldn't need any cleaning.
- Aluminum: These samples can be made by cutting small strips out of an empty beverage can and then cleaning one surface to expose fresh metal.

Part II. Preparing the Wellplate

You will use all 24 wells of your wellplate for this investigation, and it is important to label your wells to keep track of which acid is in each one. You will fill the wells with the acids, and then drop your test objects into each well to observe the reaction.

1. Place your clean, dry wellplate on a white sheet of paper. Label the rows and columns on the paper according to the diagram in Figure 19.1 (next page). It will be helpful to have a dark surface or paper to use instead of the white paper for tests that can be seen better when viewed against a dark background.

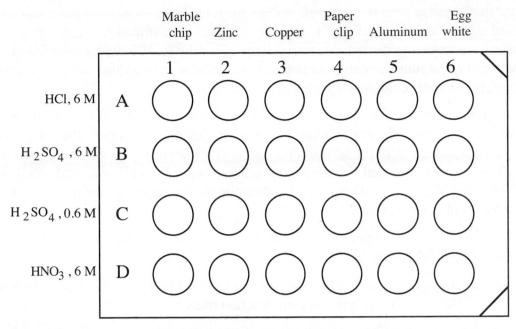

Figure 19.1. Diagram of the wellplate setup

2. Add one dropper (about 20-30 drops, or 1-1.5 mL) of 6 M hydrochloric acid (HCl) to each of the wells in Row A. The wells should be 1/4 to 1/2 filled.

     CAUTION! Be careful not to spill acids on your skin, clothing, books or papers. If you do spill any acid on your skin, wash it off promptly with large amounts of water. Continue to run water on the affected area for several minutes. Notify your instructor immediately in the event of a spill on any person or workspace.

3. Similarly, add one dropper of 6 M sulfuric acid (H_2SO_4) to each well in Row B.

4. Row C can be filled with a 10-fold dilution of the 6 M sulfuric acid. One way to do this is to put 18 drops of water in each well and then add 2 drops of 6 M H_2SO_4 to each well. Stir gently to mix.

5. Fill all wells in Row D with 6 M nitric acid (HNO_3).

6. Prepare a paper towel next to the wellplate and a plastic wash bottle filled with water. Lay test objects on the towel after they have been used.

Part III. Investigating Reactions

Note: This investigation involves many tests and observations. Careful observation is essential because the reactions of acids are so varied. *It is important to record your observations as you make them, rather than waiting until the end.*

 CAUTION! Nitric acid reacts with most metals to produce poisonous gases. Therefore, ALL of the tests with metals and nitric acid MUST be done in a fume hood. You can do the entire investigation in the fume hood if space is available.

1. Put a small marble chip in each well in Column 1. Observe carefully what happens and record all of your observations. Do you see evidence of chemical reactions? Pay particular attention to differences between the acids and to differences in reaction rates. (**Optional extension:** Test the reaction of these acids with two other familiar forms of calcium carbonate: chalk and eggshell.)

2. Put zinc strips into the HCl and H_2SO_4 wells in column 2 (but *not* into the nitric acid well unless you are doing the entire investigation in the fume hood). Try to leave a portion of the metal sticking out of the acid so that you have a "handle" with which to remove it when the test is completed. Observe the reaction of the zinc with each acid solution and record your observations. Pay particular attention to differences in the reactions and to differences in reaction rates.

3. Repeat step 2 with the copper, nickel and aluminum samples, one at a time, carefully recording your observations.

4. Put several drops of egg white into the wells in Column 6. Record your observations and note any differences in the reactions and reaction rates.

5. Finally, if you have not been working in the fume hood, take your wellplate and fresh metal samples to the fume hood. Place the metal samples into the appropriate nitric acid wells in Row D. Carefully record your observations as you did for the other samples.

 CAUTION! Do not put your face close to the wellplate when making your observations because a poisonous gas may be produced. When the reactions are finished, either leave your wellplate in the fume hood until the evolution of gas has stopped, or lift the pieces of metal out of the nitric acid solution before removing the wellplate from the hood.

Clean-up

Carefully remove any pieces of marble or metal from your wellplate and put them in appropriate containers. Empty the liquid contents of your wellplate into the waste container provided and follow any other instructions for disposal and cleaning. Rinse the wellplate several times with tap water and allow it to drain.

Analyzing Evidence

Look carefully at your observation table.

1. Which materials react with acids?
2. Did the reactions all appear to be the same, or are there differences?
3. Can you discern any patterns in the reactivity?

Interpreting Evidence

1. Marble is calcium carbonate, $CaCO_3$. Its reaction with hydrochloric acid is
$$CaCO_3 + 2\,HCl \rightarrow CaCl_2 + CO_2 + H_2O$$
What was the gas produced when marble reacted with HCl? Write chemical equations for the reactions of marble with the other two acids? Do these reactions have anything in common? In light of your observations, speculate on the damage to marble statues and buildings caused by acid rain.

2. The gas produced in the reactions of metals with acids is H_2. Write a chemical equation for the reaction of zinc (Zn) with HCl. One product is zinc chloride, $ZnCl_2$.

3. What generalization can you make about the effect of changing the concentration of an acid? Suppose the acid concentrations were reduced 100-fold, 1000-fold, or even more?

Making Claims

What can you claim about the reactions of acids with different substances? Include specific examples from your investigation data.

Reflecting on the Investigation

1. The reactivity of marble with sulfuric acid seems curiously out of line with all of the other observations. If you observed carefully, you may have seen a brief burst of gas bubbles, which quickly stopped. This is because one product of the reaction, calcium sulfate, $CaSO_4$, is very insoluble and provides a protective coating on the marble, slowing the reaction dramatically. Does this mean that acid rain containing sulfuric acid should have little or no effect on statues and buildings made of marble? Explain your answer.

2. Iron reacts readily with acid, but is also an important structural material for buildings. Based on your observations, how effective are zinc or nickel coatings as a protection for iron against acid rain?

3. Copper has been important since ancient times as a building material used on the roofs of buildings. On the basis of your observations, is copper a good material for this purpose? (Note: You may have observed a green color on the roofs of public buildings or monuments such as the Statue of Liberty. This is a form of copper carbonate, formed by slow reaction of copper with nitric acid plus carbon dioxide from the atmosphere.)

4. What differences did you observe for reactions of aluminum compared to the other metals? Aluminum easily forms a very unreactive oxide coating. How does this account for the observed differences in aluminum's reactivity?

5. Stomach acid is approximately 0.16 M HCl. Based on your observations of the reaction of acids with egg white, what effects might stomach acid have on any protein you eat? Would you expect the reaction of stomach acid with protein or food to be faster or slower than the reaction you observed between egg white and 6 M HCl?

Notes

Characterizing Acidic and Basic Materials

Asking Questions

- What common substances are acidic or basic?
- How can acidity and basicity be measured?
- What are the benefits and challenges of measuring pH with a calibrated pH meter?

Preparing to Investigate

The acidity or basicity of substances can be determined by measuring their **pH**. The pH scale is a convenient way of expressing hydrogen ion concentration in solutions. Chapter 6 in *Chemistry in Context* discusses how to measure acidity and some of the consequences of acidification of rain and ocean water. pH is defined as the negative of the logarithm of the hydrogen ion concentration, $[H^+]$, in molarity (mol/L). Equation 1 puts this in mathematical form.

$$pH = -\log [H^+] \qquad \textbf{(1)}$$

Modern pH meters are easy to use and generally reliable, but their sensitivity means that they are highly susceptible to a variety of interferences. If you use an electronic pH meter to record your pH values, you should carefully follow the directions for using it in order to record accurate values without damaging the instrument. Detailed instructions on using a pH meter can be found in the *Laboratory Methods* section of this laboratory manual, and your instructor may provide additional guidelines for the use and care of your specific pH meter. Alternatively, your instructor may have you measure pH using paper test strips that have been impregnated with dyes that change color depending on pH. This method is less accurate, but can be used.

Making Predictions

After reading *Gathering Evidence*, prepare a data table to record your predictions and results. Predict whether the substances you will test are acidic (pH < 7) or basic (pH > 7).

Gathering Evidence

Overview of the Investigation

1. Learn to operate a pH meter and calibrate the meter.
2. Measure the pH of pure acid and base solutions.
3. Measure the pH of various foods and household products.
4. Measure the pH of tap water, rainwater, and surface water.
5. Measure the pH of water containing two atmospheric gases: CO_2 and (optional) SO_2

Part I. Operating and Calibrating the pH Meter

Read the instructions for using a pH meter in the *Laboratory Methods* section and familiarize yourself with the parts and controls of the pH meter you will be using. Calibrate the pH meter using reference solutions of pH 7 and pH 4. Before each measurement you should wash the electrode with pure water and gently blot it dry with a soft tissue. The electrode should always remain submerged in water or another liquid to ensure that it never dries out. The electrodes must be handled with care because they are both fragile and expensive.

Part II. Measuring the pH of Chemical Solutions of Known Concentration

When measuring pH, you should start with the *least* acidic solution and proceed to the *most* acidic solution. Follow the instructions given in the *Laboratory Methods* section for measuring the pH of a solution. Remember to wash the electrode thoroughly before each measurement with pure water and gently blot it dry with a tissue. Never allow the electrode to dry out.

1. Calculate the pH of a 0.0001 M solution of HCl, using Equation 1. Then, measure the pH.

2. Repeat Step 1 for solutions of 0.001 M and 0.01 M HCl, and for 0.001 M NaOH. Be sure to calculate your pH *before* doing the measurement.

Part III. Measuring the pH of Foods and Household Substances

Now that you know how to measure pH, you should measure the pH of some common substances. Samples may be provided by the instructor or brought in by students. In each case, put a small amount of the substance in one of the wells of a plastic wellplate or in a small test tube or beaker. Use only enough so that the bulb end of the electrode is immersed. Check with your instructor if you have questions. Remember to rinse the electrode thoroughly and blot it dry before each measurement.

Below are some ideas for substances to test. You should measure the pH of 6-8 substances or the number specified by your instructor.

- vinegar
- lemon juice
- fruit juices

- soft drinks
- coffee or tea

The following substances must be diluted before testing the pH. To do this, mix a few drops of liquid or a very small scoop of solid with pure water.

- liquid dish detergent
- dishwasher detergent
- laundry detergent
- shampoo or hand soap

- household ammonia
- liquid laundry bleach
- baking soda (sodium bicarbonate)
- drain cleaner (***Caution:*** *see note below*)

 CAUTION! Drain cleaners can be extremely caustic. They are designed to dissolve hair and other debris. Be especially cautious about getting any drain cleaner on your skin, and always wear your safety glasses. *In case of any skin contact, wash with copious amounts of water and notify the instructor immediately.*

Part IV. Measuring the pH of Tap Water, Rainwater, and Surface Water

1. Measure the pH of tap water from the laboratory sink, a drinking fountain, or elsewhere on campus.

2. The pH of rain or snow is particularly interesting because of concern about "acid rain." Your instructor may provide samples that have been collected recently or you can collect your own sample using the procedure described in Investigation 21. Remember that rain is nearly pure water and any acid present will be dilute, which can make pH changes difficult to measure.

3. It is instructive to compare the pH of rain to the pH of river water. If river samples are available in the lab, check the pH.

4. If water from a swimming pool is available, check its pH.

Part V. Measuring the pH of Water Containing CO_2 or SO_2

1. Test the pH of water that is saturated with carbon dioxide. You can use a sample of seltzer water that has been allowed to go "flat", or you can generate carbon dioxide as described in Investigation 1. Test the pH of the water containing CO_2.

2. Test the pH of water that is saturated with air. Remember that CO_2 is only a very small fraction of the gases that make up air.

3. Test the pH of water that is saturated with SO_2.

 CAUTION! Sulfur dioxide is a very toxic gas, and some individuals have a serious allergic reaction to SO_2. This test should be done *only* in a good fume hood and should not be attempted by individuals who are allergic to SO_2.

   a. Make SO_2 in a resealable zipper bag. Fill a plastic pipet with 6 M sulfuric acid and place it in a quart-size plastic zipper bag. Add about 1-2 grams of sodium sulfite (Na_2SO_3) to the bag. Close the bag securely, excluding the air, and then squeeze the pipet so that the chemicals will mix and react.
   b. Saturate water with SO_2. With the bag lying on the lab bench, <u>very cautiously</u> open a corner of the bag, squeeze the air out of a dry plastic pipet, insert it into the bag and fill the pipet with SO_2. <u>Reseal the bag immediately</u>. Slowly bubble this gas into pure water in a test tube or wellplate, and then test the pH of the water.
   c. Dispose of your SO_2 sample into a designated waste container in the fume hood.

Clean-up

Dispose of your samples in appropriate waste containers as instructed. Clean your pH electrode and store it in water or another solution as designated by your instructor.

Analyzing Evidence

Look carefully at your results table.

1. In Part II, did your calculated pH values match your measured pH values?

2. In Parts III and IV, did you correctly predict which substances were acidic and which were basic? How well did your predictions match up with your pH measurements?

3. In Part V, were your solutions acidic or basic?

Interpreting Evidence

1. Can you propose a general rule for how the pH should change (up or down and by how much) when the acid concentration increases by a factor of 10? For example, when HCl concentrations change from 0.0001 M to 0.001 M to 0.01 M, the acid concentrations increase by a factor of 10 each time.

2. Were tap water and surface water acidic or alkaline? What substances may account for this?

3. Is the pH of water from a river different from the pH of collected rain? If so, suggest an explanation.

4. Is the pH of swimming pool water different than tap water? If so, suggest an explanation.

5. Were the measured pH values of CO_2 and SO_2 in water what you expected? Write chemical equations for the reaction of each gas with water.

Making Claims

What can you claim about the pH of different substances?

Reflecting on the Investigation

1. Explain why pH is a useful way of describing acid and base solutions over a very wide range of concentrations.

2. Can you draw any conclusions about the pH of different categories of common substances? What is the general pH of beverages? Of soaps and detergents? Of substances that normally touch human skin?

3. In Part IV, was the rain acidic or alkaline? Since rain is formed by evaporation and condensation in clouds, why is the pH not the same as pure water?

4. In your own words, describe how (a) increased concentrations of SO_2 lead to acid rain; or (b) increased concentrations of CO_2 lead to ocean acidification.

Acid Rain

Asking Questions

- Does acid rain fall where you live?
- What is the normal pH of rain?
- What substances in rain are responsible for making it acidic?
- What are some of the consequences to infrastructure and ecosystems of acid rain?

Preparing to Investigate

The pH scale is a convenient way of expressing the hydrogen ion concentration in solutions. pH is defined as the negative of the logarithm of the hydrogen ion molarity $[H^+]$. Equation 1 puts this in mathematical form.

$$pH = -\log[H^+] \tag{1}$$

All water samples contain some hydrogen ions. Acidic solutions have a higher concentration of hydrogen ions than do alkaline solutions. In water at 25°C, the product of the molarities of H^+ and OH^- is always 1.0×10^{-14}. In pure water, the concentrations of H^+ and OH^- are equal to each other, both 1.0×10^{-7}, so the pH is 7. In acidic solutions, the H^+ ion is present in excess compared to OH^-; the concentration will be higher than 1.0×10^{-7}, and thus the pH will be less than 7. Chapter 6 in *Chemistry in Context* discusses pH in greater detail.

Pure rainwater has a pH of approximately 5.6, due to the presence of carbon dioxide. CO_2 reacts with water to produce carbonic acid, which then dissociates into hydrogen ions (H^+) and bicarbonate (HCO_3^-).

$$CO_2(g) + H_2O \rightarrow H_2CO_3 \rightarrow H^+ + HCO_3^-$$

When rain has a pH below 5.6, this means that other acids must be present. Rain can be classified as follows:

Rain classification	pH
Pure rain	5.6
Slightly acidic	5.0-5.6
Moderately acidic	4.5-5.0
Highly acidic	4.0-4.5
Extremely acidic	Below 4.0

The average pH of rain varies across the world. To learn more, you may wish to visit the U.S. Environmental Protection Agency website and view the acid rain maps that are available there. If you live outside the U.S., you can do a web search for similar maps of your country.

Making Predictions

After reading *Gathering Evidence*, prepare a data table to record your predictions and pH measurements. Predict the relative acidity of your rain samples. How do you think acidity will vary based on where and when they were collected?

Gathering Evidence

Overview of the Investigation

1. Prepare sample containers for collecting rain samples.
2. Select sites for the collection containers and collect the samples.
3. Prepare the samples for pH measurement.
4. Calibrate the pH meter and measure the pH of the rain samples.

Part I. Preparing Sample Containers

Samples should be collected in plastic containers, not glass, and it is important that the containers be thoroughly cleaned in advance. The recommended cleaning procedure is to first rinse each container with 6 M hydrochloric acid, then five times with tap water and finally five rinses with distilled or deionized water. Once cleaned and dried, the containers should be capped or covered to keep them clean.

 CAUTION! 6 M HCl is a corrosive liquid. Wear gloves and a lab apron or coat along with your safety glasses.

Part II. Collecting, Handling and Storing Rain Samples

Your instructor may assign you a location for collecting rainwater samples, or your class may discuss and decide collectively on the best collection locations for answering the questions you have. If your bottles have a large opening, they can be set out to collect rain. Alternatively, in order to obtain larger samples, you may need to use a large plastic funnel (which should also be cleaned in advance). With a funnel, the bottle may need to be supported in an upright position by, for example, placing it inside a metal can that is nailed to a post. The collection container should be placed in an unobstructed location where it will not be disturbed and ideally will be several feet above ground level. This will minimize contamination of the sample by dirt and dust.

A single sample can be collected for an entire rain event. However, the pH of rain can change significantly during the course of a rain event, so it can be interesting to collect samples at intervals during the event. For rain lasting one or several days, the container might be changed every few hours or twice a day. Alternatively, the container might be changed every 15 minutes during an afternoon thunderstorm. However you collect your samples, it is important to label each container clearly with the date, time of day, total sampling time, and location where it was collected.

Samples should be brought into the laboratory for measurement as soon as possible, and filtered into another clean bottle to remove any dust and debris that might react with the acids in the rain. If filtration and pH measurement cannot be done immediately after collection, samples should be stored in the refrigerator until they can be brought to the lab.

Part III. Measuring pH of Rain Samples

The pH of the rain samples will be measured using a pH meter. Operation and care of the pH meter is described in the *Laboratory Methods* section of this lab manual. The pH meter must first be calibrated using standard buffer solutions of pH 7 and pH 4. When calibrating, gently stir or swirl the buffer solution with the electrode for 30 seconds, and then let the electrode stand undisturbed in the buffer for 30 seconds. Adjust the meter as needed to the correct pH.

Rinse the electrode thoroughly with distilled or deionized water between each sample, and gently blot it dry with a tissue. Insert the electrode into a clean beaker containing 10-20 mL of the rainwater sample. Stir or swirl the solution to ensure homogenous contact with the electrodes, and then allow the solution to settle for approximately 30 seconds. Record the pH after the reading has stabilized. Record all of your pH values in your data table.

Clean-up

Dispose of your samples in appropriate waste containers as instructed. Clean your pH electrode and leave it soaking in water. Clean all glassware by first rinsing with tap water and then with distilled or deionized water.

Analyzing Evidence

1. Describe the collection site(s) for your samples, and briefly explain the predictions you made about the relative acidity of the samples.

2. Did the pH of your samples match what you predicted? Explain.

3. Collect the class data and classify each of the samples according to the acid rain scale in the table at the beginning of the investigation.

Interpreting Evidence

1. If several samples from one rain event are available, how did the pH change during the event? Try to provide an explanation for any changes you see.

2. If the samples were collected at different locations, are there any pH differences between locations? If so, how can you explain them?

Making Claims

What can you claim about how location and timing affect the pH of rain?

Reflecting on the Investigation

1. Based on your measurements, do you live in an area affected by acid rain? If so, have you noticed any of the effects of acid rain that are described in your textbook?

2. Sometimes, rain samples are found to be not acidic at all, and instead are alkaline with a pH greater than 7. What might account for this? Formulate a hypothesis. What chemical tests or investigations could be performed to support or refute your hypothesis?

Investigating Solubility

Asking Questions

- How do scientists develop a procedure for measuring properties of materials in the laboratory?
- What data is important when defending a method for solving a problem?

Preparing to Investigate

This laboratory exercise is a departure from the usual investigation. It is an attempt to simulate the kind of problem solving that takes place in a scientific laboratory. There are no instructions or procedures. There is simply a problem to solve, which requires reasoning skills and the application of knowledge that has been previously acquired. You will work in a team of three or four members.

A variety of materials and equipment will be available that you can use to solve the problem. At a minimum, your "lab" should have graduated cylinders, burets, beakers, flasks, plasticware, stirrers, test tubes, plastic pipets, and an analytical balance. You may wish to review the *Laboratory Methods* section at the beginning of the laboratory manual for a review on how to use some of this equipment.

Making Predictions

First, assign roles to each person in your group, including someone to record your ideas, procedures and results. After reading *Gathering Evidence*, develop a method to solve the problem, and then perform your investigation and record your results. Be prepared to compare your method and results with those of the other teams in the class.

Gathering Evidence

Your task is to determine the solubilities of the following compounds in water at room temperature:

- calcium sulfate – $CaSO_4$
- potassium aluminum sulfate – $KAl(SO_4)_2$
- potassium nitrate – KNO_3
- ammonium nitrate – NH_4NO_3
- copper nitrate – $Cu(NO_3)_2$
- sodium chloride – $NaCl$

Solubility is defined as the maximum mass of solid that can dissolve in a given volume of water. It is usually reported as grams of solid per mL of water, or grams of solid per 100 mL of water. Think carefully about how the procedure you design will help you measure this information.

All of these compounds, except the copper salt, are likely to be found in drinking water supplies since they occur naturally in rocks and soil or are used as components in fertilizers. Copper nitrate is included because copper salts are often used to reduce algae in ponds and swimming pools and so is also found in the water supply.

In the interest of cost and convenience, you may use no more than 2 g of each compound. Your group can use graduated cylinders, burets, beakers, flasks, plasticware, stirrers, test tubes, plastic pipets, and an analytical balance.

Your team should be prepared to defend your answer and the method you used to obtain it. Therefore, it is important for your team to keep a complete record of everything you do and the numerical data you obtain. It also is important that you present the data to provide evidence that your method solves the presented problem. Your grade for this investigation will not depend on getting the "correct answer", but rather on the quality of the procedure you develop and the care with which you carry out your investigation.

Analyzing Evidence

After you perform your investigation, rank the compounds from least soluble to most soluble.

Interpreting Evidence

1. Compare your methods and results to those of your classmates. How well did your procedure work? Can you think of any modifications or improvements you would implement if you were to do this type of investigation again?

2. Compare your results to the known solubilities of the compounds. Did you rank the solubilities correctly?

Making Claims

What can you claim about the process of science?

Reflecting on the Investigation

What insights do you have into how scientists design investigations and work collaboratively?

Measuring Radon in Air

Asking Questions

- What is the relationship between alpha particles and helium nuclei?

- What property of radon gas can be used to detect it?

- What factors impact radon levels in buildings?

- Why is it important to measure radon concentrations in our living environment?

- How can high levels of radon be mitigated?

Preparing to Investigate

Radon is a radioactive gas that is a potentially serious indoor health hazard in some geographic areas (see Chapter 7 of *Chemistry in Context*). Radon-222 (the longest lived isotope of radon) is an alpha-emitter with a half-life of 3.8 days. The atomic number of radon is 86, which places it in Group 8A at the far right-hand side of the periodic table. The elements in Group 8A are chemically inert gases that generally do not form molecules.

As shown in Figure 7.12 in *Chemistry in Context*, radon is formed as an intermediate product in the radioactive decay of uranium-238. Uranium is widely distributed on the surface of our planet. For example, granite rocks contain about 4 ppm of uranium. When radon is produced from uranium, it passes through fissures in rocks and soils without reacting and enters homes through cracks or other holes in the foundations. The amount of radon found in homes varies from one part of the country to another and is typically more serious in houses with basements. National and state maps of predicted radon concentrations in homes can be found at the following EPA website: http://www.epa.gov/radon/zonemap.html.

In this investigation, you will use a simple radon detector to measure the radon concentration in a location of your own choosing. A sampling period of at least 4 weeks is necessary in order to obtain reliable data.

The disk in the detector consists of a high-clarity polymer often used in eyeglasses that is known as CR-39. The full chemical name for CR-39 is poly[ethylene glycol bis(allyl carbonate)]. The alpha particles (see Table 7.1 in the text) emitted from radon are ionizing radiation. When they penetrate CR-39, they cause damage in the plastic, probably due to disruption of the polymer chain along the path of penetration. Although the damage is not visible to the eye, treating the disk with sodium hydroxide, NaOH, reveals it. Sodium hydroxide etches the sample in the damaged regions. The etched regions show up as "tracks" when viewed under a microscope and the number of tracks in a given area of the disk can be used to estimate the radon level at the sampled location. Although beta and gamma radiation also penetrate the plastic, these rays mostly pass through without causing damage.

Making Predictions

Look at the maps on the EPA website cited in the introduction. Do you predict that radon will be found in the area where you placed your sampling disk?

Gathering Evidence

Overview of the Investigation

1. Place a detector disk in an undisturbed location for at least 4 weeks.
2. Etch the disk to produce visible alpha tracks.
3. Count the tracks under a microscope.
4. Calculate the concentration of radon in air by comparison with a control disk.

Part I. Sampling the Air

Start the air sampling early in the semester as directed by your instructor. A long sampling period, preferably for *at least* a month, is necessary. Select a location where the detector can remain undisturbed and where you can retrieve the detector prior to the scheduled lab time for this study. (*Plan this carefully, perhaps adding a label to explain what the item is. Custodians or other persons may unknowingly remove the disk when cleaning up an area!*) The class results will be more interesting if students select many different locations, including on and off campus, inside buildings and outdoors, various floor levels in buildings, and with different ventilation or air circulation systems. One possibility is to use your own home, either taking the detector device there yourself or mailing it to a relative with instructions. In some parts of the country, basements of houses are particularly prone to high levels of radon.

A. Materials

1. Piece of CR-39.

2. Small plastic cup, preferably with lid.

3. Rectangle of cardstock (size depends on the cup being used).

4. Piece of transparent adhesive tape.

5. Piece of thin, single-layer toilet or tissue paper (to protect the CR-39 from dust).

B. Procedure for Use at Sampling Location

1. If the cup has a lid, cut a large circular hole in the lid.

2. Remove the plastic film from the marked side of the CR-39.

3. Make a loop of tape with the sticky side out. Stick the tape to the side of the CR-39 that still has the plastic film coating in place. Stick the CR-39 to the middle of the piece of cardstock.

4. Bend the cardstock into an inverted U that will support the CR-39 near the top of the cup. Put the cardstock plus CR-39 into the plastic cup. Cover the opening of the plastic cup

with the toilet or tissue paper, and then snap the lid onto the cup. (If the cup does not have a snap-on lid, use a rubber band to hold the tissue paper in place.)

5. Place the detector in the chosen location, and record the location and start date and time. **Note:** It is important that you accurately record the location for the detector disk (describing it in detail) and the start and stop times (dates and time of day).

6. At the end of the air-sampling period, record the ending date and time. Store the card with the attached disk in a labeled envelope until time for the laboratory work.

Part II. Etching the Disk

Remember to bring your exposed disk to your laboratory class!

1. Remove the disk from the cardstock. You already peeled off one of the polyethylene films from the disk. Now peel off the other (from the back of the disk). Gently slip the ring of an "etch clamp" over the top of the disk. *CAUTION: The disk is fragile and can break easily.* Fasten the ring onto a paper clip that has been bent so there is a hook at each end (see *Figure 23.1*).

2. Hook the disk inside a test tube and use a pipet to add enough 6 M sodium hydroxide solution (NaOH) just to cover the disk in the tube. *CAUTION: Do not submerge any more of the paper clip than is necessary because the coating on the paper clip will react with NaOH..*

CAUTION! Hot 6 M NaOH is extremely corrosive to skin and to clothing. *Wear goggles at all times. Rubber gloves are strongly recommended.*

3. Heat a beaker of water on a hot plate to a boil. Carefully place the test tube containing the disk and NaOH solution in the boiling water bath and heat it for 40 minutes. At the end of 40 minutes, remove the disk and rinse thoroughly with lots of tap water.

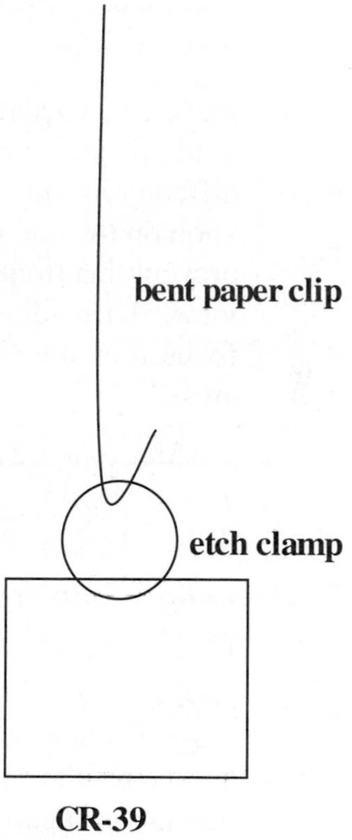

bent paper clip

etch clamp

CR-39

**Figure 23.1
Investigation set-up
for etching**

Clean-up

When finished, do NOT pour the NaOH solution down the drain. Pour it into an approved container provided by the instructor.

Part III. Counting Alpha Tracks

The measuring process will be completed using a microscope. Your instructor or teaching assistant will show you how to operate the particular microscope that you will be using.

A. Calibrating and Focusing the Microscope

1. Determine the area (in square centimeters) of the field of view of your microscope at low power (10x) by looking through the microscope at a ruler. Count the number of millimeter divisions on the ruler that you can see across the widest portion of the field of view and record the diameter in millimeters. Divide by 10 to find the diameter in centimeters. The radius, r, is half of the diameter, and the area, A, can be calculated by using the formula $A = \pi r^2$.

2. Place an etched disk on a microscope slide under the microscope. Then adjust the focus until you are sure you have focused on the etched disk. The glass slide and disk are at different heights, so they will come into focus at different times as you adjust the focus knob on the microscope. It is easy to mistake the microscope slide for the disk. To prevent this from occurring, move the microscope slide so that the edge of the disk is in view. Then adjust the focus until the disk comes into sharp focus. Once you have focused on the disk, you can move the slide around to observe the tracks in different areas.

3. Practice looking at alpha tracks on a previously exposed and etched disk. This may be a disk donated by a student in a previous class, or it may be one prepared by your instructor using a radioactive source (in which case, your own disk likely will have fewer tracks).

B. Counting Alpha Tracks

1. Look at the two pictures of alpha tracks at different magnifications (*Figure 23.2, next page*). The left photo (*Figure 23.2 A*) looks closer to what you will see. At higher magnification, you can see tracks of different shapes (*Figure 23.2 B*). The various shapes are the result of how the alpha particles entered the solid. The circular-shaped tracks are due to alpha particles that entered straight (perpendicular to the disk), and the more teardrop-shaped tracks are due to alpha particles that entered at an angle.

2. Place your disk on a microscope slide. Look through the microscope and adjust the focus until you are sure you are focused on the etched disk. Again, be sure not to mistake the microscope slide for the disk. Once you have focused on the disk, you can move the slide around to observe the tracks in different areas.

3. Count the number of tracks in your field of view and record your data on your data sheet. Move the disk to a new field of view and count and record the number of tracks in the new field of view. Repeat until you have collected and recorded data from *10 different fields of view*.

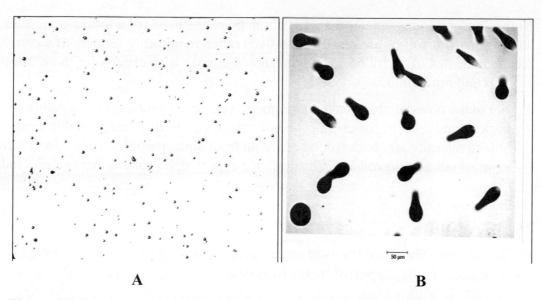

A   **B**

Figure 23.2 Appearance of alpha tracks on an etched CR-39 disk that had been exposed to 9 pCi/L for five days. (A) At 20X magnification and (B) At high magnification, showing different shapes of tracks.

Analyzing Evidence

1. Average all your measurements to determine the average number of tracks in a field of view for your disk.

2. Calculate the average number of tracks per square centimeter by dividing the average number of tracks by the area of your field of view (Calibrating, Step 1).

3. Calculate the number of tracks per square centimeter per day of exposure (including any fraction of a day). See your instructor if you need help with this calculation.

4. Control plastic disks that were sent to a radon facility in which the radon level was known to be 370 picocuries per liter were found to exhibit 2,370 tracks/cm^2/day by this etching and counting technique. Use this relationship between picocuries per liter and tracks per square centimeter per day to calculate the radon level for *your* sample in picocuries per liter of air.

Interpreting Evidence

Answer these questions in paragraph form and include the answers in your written report.

1. You are comparing your measurements to those in picocuries. What does pico mean? What does a curie measure? See your textbook if needed.

2. How does your sample compare with the 4-pCi/L guidelines set by the Environmental Protection Agency?

Note: Concentrations greater than 4 pCi/L indicate the need for a follow-up test. You can do another one yourself or use an EPA-approved radon monitoring service. If a concentration greater than 4 pCi/L is found again, you should consult with experts for more detailed analysis and mitigation.

3. For many types of chemical measurement, you need to subtract out a naturally occurring background level of the chemical being measured. In this case, you don't expect to be able to measure any background radiation from alpha particles, so we don't have to subtract out a background. Why don't we expect alpha particle background radiation?

Making Claims

What can you claim about the levels of radon in different locations? What data suggest these claims? Are your claims supported by the EPA data?

Reflecting on the Investigation

Answer each of the following questions in paragraph form in your written report.

1. Is your calculated radon concentration consistent with your prediction? Give some potential reasons why it is or is not as you expected.

2. Consider the following situations and determine if radon is likely to be a problem for the occupants of the buildings. Explain why or why not.

   a. A beachside home in Mayagüez, Puerto Rico, a university town on the west coast of this volcanic Caribbean island. The home is built on a concrete slab and has screen windows.

   b. A ten-story office building in chilly Duluth, Minnesota. The building is well insulated and sealed against the winter arctic blasts.

   c. A two-story farmhouse in rural Nebraska. The basement has a dirt floor with underlying limestone rock.

3. The following health advisory was issued by the surgeon general:

   "Indoor radon gas is a national health problem. Radon causes thousands of deaths each year. Millions of homes have elevated radon levels. Homes should be tested for radon. When elevated levels are confirmed, the problem should be corrected."

   Consult your textbook and list 3 things that can be done to mitigate high radon levels. Explain briefly how each of the actions you list will help solve the problem.

4. Visit www.epa.gov/radon/zonemap.html. Find your home and college on the maps. What are the expected levels of radon in your area? Should your family be concerned about radon?

Exploring Electrochemistry

Asking Questions

- How can we get electricity from chemical reactions?

- How can electricity contribute to a chemical reaction?

- What is the relationship between a galvanic cell and an electrolytic cell?

- How do electrolytes influence electrolysis?

Preparing to Investigate

Any spontaneous chemical reaction that involves the transfer of electrons from one atom or molecule to another can be used as the basis for an electrochemical cell. Consider, for example, the reaction between zinc metal and copper ions in water:

$$Zn(s) + Cu^{2+}(aq) \rightarrow Zn^{2+}(aq) + Cu(s)$$

If zinc metal, $Zn_{(s)}$, is added to a water solution containing copper ions, $Cu^{2+}(aq)$, the blue color of the copper ions disappears, the reddish color of metallic copper appears, and the zinc seems to disappear. This process occurs through electron transfer. Each zinc atom loses two electrons and forms a +2 ion. Each copper ion adds two electrons and forms metallic copper. The zinc ions are colorless and soluble in the water, and the copper precipitates out. As explained in Chapter 8 in the textbook, we can break this up into two half reactions:

$$Zn(s) \rightarrow Zn^{2+}(aq) + 2 e^-$$

$$Cu^{2+}(aq) + 2 e^- \rightarrow Cu(s)$$

In order to get useful electric energy from this reaction, the two half reactions have to occur in two separate locations, but they must be connected so that the electrons can flow through an external wire. The wire can be connected to something (such as a motor, light bulb, or meter) to show that an electrical current is produced. A device with these components is called a **galvanic cell** (see Your Turn 8.4 in *Chemistry in Context*), although the nonscientific and slightly incorrect name *battery* is often used. In this investigation, you will investigate how to assemble several working galvanic cells and use a voltmeter to measure the voltages produced.

Rather than allowing a chemical reaction to produce electric energy (in the form of a flowing electron current), the reverse is also possible. In an **electrolytic cell** (see Section 8.6 and Figure 8.15 in *Chemistry in Context*, electric energy from an external source, such as a commercial "battery," is used to force a chemical reaction to go backwards, i.e., in the non-spontaneous direction.

Both types of cells have many practical applications. Everyone is familiar with electrochemical cells ("batteries") ranging in size from the tiny cells in hearing aids or watches to large automobile or truck batteries. Electrolytic cells are widely used in chemical manufacturing to produce chemicals that otherwise would be difficult to make.

Making Predictions

- After reading *Gathering Evidence*, locate the three metals used in today's investigation on the periodic table. What are their atomic numbers? Based on what you know about periodicity, how do you think these metals will react? Rank the galvanic cells (Zn/Ag, Zn/Cu, Cu/Ag) by predicted voltage and identify what leads you to expect this ranking.

- Prepare a data sheet that includes room for your predicted voltage rankings and reason for the prediction, your voltage data for the three cells, a labeled drawing of your galvanic cell, the measured voltage ranking for the cells, and observations for both the galvanic cells and electrolytic cells.

Gathering Evidence

Overview of the Investigation
1. Assemble three galvanic cells and measure their voltages.
2. Construct and test a "citrus cell."
3. Assemble a simple electrolysis unit.
4. Carry out the electrolysis of water.
5. Carry out the electrolysis of potassium iodide solution.
6. Compare the two electrolysis reactions and identify differences between the two.

Note: This investigation has two parts that can be done independently and in either order. To facilitate efficient use of equipment, your instructor may assign half of the class to start with Part I (galvanic cells) and the other half to start with Part II (electrolytic cells).

Part I. Generating Electricity from Chemical Reactions

You will set up three different galvanic cells:

1. $Zn \rightarrow Zn^{2+} + 2\ e^-$ with $2\ Ag^+ + 2\ e^- \rightarrow 2\ Ag$

2. $Zn \rightarrow Zn^{2+} + 2\ e^-$ with $Cu^{2+} + 2\ e^- \rightarrow Cu$

3. $Cu \rightarrow Cu^{2+} + 2\ e^-$ with $2\ Ag^+ + 2\ e^- \rightarrow 2\ Ag$

In order not to waste chemicals, these will be set up at a small scale using a piece of filter paper placed on a watch glass.

1. Obtain a watch glass and a piece of filter paper of the same size. On the filter paper, draw three small circles, labeling them Ag^+, Cu^{2+}, and Zn^{2+} as shown in *Figure 24.1*. Trace a trail from each circle to the center of the paper. Use scissors to cut out wedges of paper between the circles. Place the filter paper on the watch glass.

2. Obtain small pieces of each of the three metals: Ag, Cu, and Zn. Sand both sides of each piece of metal with sandpaper until it is clean and shiny.

3. Place 3 drops each of 0.1 M solutions containing Cu^{2+}, Zn^{2+}, and Ag^+ on the corresponding circles.

 NOTE: Although it is not harmful, silver nitrate will produce a black stain on skin. Try to avoid getting this solution on your skin.

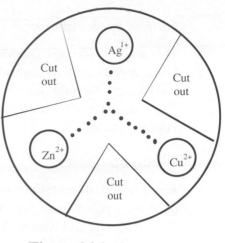

Figure 24.1

Galvanic cell assembly

4. Place each piece of metal on the spot of its corresponding ion (e.g., the Zn goes on the Zn^{2+} spot). *Make sure that the tops of the metal pieces are kept dry.*

5. Attempt to measure the voltage for the zinc and copper combination as follows. Obtain a voltmeter and learn how it operates. Turn it on and select a range of about 1 or 2 volts. (Some meters automatically select the appropriate range.) Check to be sure that the meter reads zero volts when the two probes are touched together. Then touch the two metal probes from the meter to the pieces of copper and zinc and observe the voltage.

6. It is probable that you observed zero volts (or very close to zero), and you should convince yourself that there is a missing link that prevents a complete cycle or circuit. To complete the circuit, fill a small plastic pipet with 1 M KNO_3 solution and carefully dribble a small amount of solution along the lines connecting the circles to the middle of the paper. Be sure there is a continuous trail of KNO_3 connecting the circles.

7. Test the voltage again by touching the probes to the Cu and Zn metals. If a positive voltage is displayed, measure the voltage for 5 seconds and record the reading. If a negative voltage is displayed, reverse the leads, measure for 5 seconds, and record the reading. Also record which metal the red probe is touching when the meter shows a positive voltage.

8. In a similar fashion, measure the voltage for the other two possible combinations: Cu + Ag and Zn + Ag. Record the voltages and which metal the red probe is touching when the meter shows a positive voltage.

9. The Citrus Cell. The metal ions that were supplied by solution in the last part of this investigation are available from many natural sources, e.g., very hard water or fruits and vegetables. To illustrate this, you can make a "battery" from a piece of citrus fruit.

   a. Choose two of the three metals (Zn, Cu, Ag).

   b. Push the metal strips through the fruit rind so that a part of each is inside the juicy portion of the fruit. The two pieces of metal must not touch but should be in the same section of the fruit.

   c. Record the voltage and which metals you used.

Further Explorations

This study could be expanded to include other metals, including iron, nickel, tin, or magnesium. Some metals such as aluminum form a very tough coating that makes this kind of measurement impractical. Certain other metals, such as sodium and potassium, are so reactive that they will immediately react with the water! Write a brief description of the exploration you plan, have your instructor check it for safety and feasibility, and, if approved, carry it out.

Clean-up

Rinse the small metal pieces with pure water and blot dry with a paper towel. Dispose of the used filter paper in an appropriate waste receptacle. Rinse off the watch glass with pure water. Return metals and watch glass to the supply location.

Part II. Promoting Chemical Reactions with Electricity

Assemble a simple electrolysis unit using two 9-volt batteries, as shown in *Figure 24.2*. (A single 9-volt battery can be used, but two batteries connected in "parallel" are preferable and will give more dramatic results.)

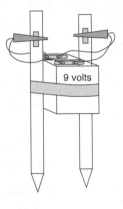

1. Tape two 9-volt batteries together.

2. Obtain two snap-on battery connectors and two alligator clips, one red and one black. Twist the ends of the two red wires together. Thread these combined wires through the hole in the handle of the red alligator clip. Bend the wire back on itself and twist to make it snug. Repeat this process for the black wires.

3. Obtain two graphite rods. (Ordinary pencils with a good point work well for this but must have a place to attach an alligator clip, either by sharpening the top end or by cutting away the wood near the middle.)

4. Tape the rods or pencils to either side of the batteries. Snap the battery connectors onto the batteries and clip one alligator clip to each graphite rod.

Figure 24.2

Electrolysis unit

A. Electrolysis of Water

Use your electrolysis unit to split water (H_2O) into hydrogen (H_2) and oxygen (O_2).

1. Fill a petri dish about half full with distilled or deionized water. Add 2 drops of phenolphthalein indicator and gently swirl to mix. This indicator is colorless in neutral or acidic solutions (excess H^+) and pink in basic solutions (excess OH^-). (Your instructor may provide a different indicator and, if so, will tell you what color changes to expect.)

2. Immerse the tips of the graphite rods and observe carefully. Do you see any evidence of a chemical reaction?

3. Add about one-half teaspoon (about 1 gram) of solid potassium nitrate, KNO_3. Jiggle the dish gently to aid in dissolving the KNO_3 crystals. KNO_3 is an electrolyte that helps the

solution conduct electricity by providing K^+ and NO_3^- ions. It *does not* participate directly in any reactions.

4. Again immerse the graphite rods in the solution. Observe carefully for a couple of minutes. You should see evidence of a chemical change. (Put the dish on a light-colored background to make the observations clearer.) Record what is happening at the electrode attached to the red wires and what is happening at the electrode attached to the black wires. Which one produces more gas bubbles? Which causes the indicator to change color?

5. Dispose of the solution in an appropriate waste container.

B. Electrolysis of a Water Solution of Potassium Iodide, $KI_{(aq)}$

1. Using the same dish from Part A, add about one-half teaspoon (about 1 gram) of solid potassium iodide, KI, to the dish.

2. Fill your Petri dish at least half full with tap water. Jiggle the dish gently to aid in dissolving the KI crystals. The KI serves two functions in this electrolysis. It serves as the electrolyte, helping the water conduct electricity, but it also will be a reactant.

3. Immerse the graphite electrodes in the solution. Record your observations, noting what is happening at or near each of the two electrodes.

4. While the electrolysis is running, add 2 drops of phenolphthalein indicator directly next to each electrode. As noted previously, this indicator is colorless in neutral or acidic solutions (excess H^+) and pink in basic solutions (excess OH^-). Record any color changes you see and note whether these were at the red or black wire electrode.

5. Dispose of the solution in an appropriate waste container and return the electrolysis unit or disassemble it if instructed to do so.

Analyzing Evidence

Record the answers to the following questions on your data sheet.

1. Sketch a diagram of your Cu/Zn cell showing the metals and solutions involved as well as the voltmeter. Show the direction of electron flow out of one metal and through the voltmeter to the other metal. (Recall that the red-wire probe was on the metal that consumes electrons.) Show the rest of the path that negative charge must follow to complete the circuit.

2. Rank the three cells in order by the voltages they generated (list the voltages too). Include in your rankings a standard Zn/MnO_2 dry cell (1.5 volts), similar to what is used in flashlights.

3. Which metals did you choose to make your citrus cell? Why? What was the voltage of this galvanic cell? How many of these would you need to get a voltage equivalent to a dry cell?

4. What two gases are the final products from the electrolysis of water? Write a balanced equation for the electrolysis of water. (Water is the only reactant, and the gases are the products.)

Interpreting Evidence

Answer each of the following questions in paragraph form in your written report.

1. How does your predicted ranking for the voltage of the three galvanic cells (Zn/Ag, Zn/Cu, Cu/Ag) compare with your data from the investigation? Explain any differences and how the investigation changed your thoughts about these cells.

2. Explain why the KNO_3 solution was needed for your galvanic cells to function.

3. The two half-reactions for the electrolysis of water are:

$$4\,H_2O(l) + 4\,e^- \rightarrow 2\,H_2(g)\ +\ 4\,OH^-(aq) \quad \text{(2 moles of gas and a basic solution formed)}$$

$$2\,H_2O(l) \rightarrow O_2(g) + 4\,e^- +\ 4\,H^+(aq) \quad \text{(1 mole of gas and an acidic solution formed)}$$

Based upon your observations, which of the two half-reactions appears to be occurring at the black-wire lead? Explain your reasoning. Which gas formed at which color electrode? Explain how you decided.

4. In light of your answer to #3 and the equations for the half-reactions, which color electrode supplied electrons from the batteries? What was the other electrode doing? Explain.

Making Claims

*   What can you claim about the influences of different metals on galvanic cells?

*   What can you claim about the influence of electrolytes and reagents on the electrolysis of water?

Reflecting on the Investigation

Answer each of the following questions in paragraph form in your written report.

1. Imagine that you are marooned on an isolated island with a few other people. After doing this investigation, what could you offer to the group that would be useful for supplying energy? Suggest something specific you could do with a few pieces of metal and food scraps or some nearby salt deposits.

2. In Part II-B, during the electrolysis of the KI solution, either H_2O or KI could react at each of the electrodes. Thus, although only one reaction will actually occur at each electrode, we must consider two possible reactions for the red electrode and two possible reactions for the black electrode. The half-reactions that could occur are the following:

At the electrode where electrons are consumed:

    Either: $4 H_2O_{(l)} + 4 e^- \rightarrow 2 H_2(g) + 4 OH^-(aq)$ (gas and basic solution formed)

    Or: $K^+(aq) + 1 e^- \rightarrow K(s)$ (metallic potassium formed)

At the electrode where electrodes are released:

    Either: $2 H_2O(l) \rightarrow O_2(g) + 4 e^- + 4 H^+(aq)$ (gas and acidic solution formed)

    Or: $2 I^-(aq) \rightarrow I_2(aq) + 2 e^-$ (yellow-brown elemental I_2 formed)

a. From your observations, what formed at the black-wire lead? Explain your reasoning.

b. From your observations, what formed at the red-wire lead? Explain your reasoning.

c. Compare your electrolysis of water to that of the KI solution. Which product from the water electrolysis was also produced in the KI electrolysis?

d. Which product from the water electrolysis was NOT produced in the KI electrolysis? What was formed instead?

e. Electrolysis will always produce the products that are easiest to make from the starting materials. What can you conclude about how easy it is to remove electrons from iodine in KI versus how easy it is to remove electrons from oxygen in H_2O?

Notes

Polymer Synthesis and Properties

Asking Questions

- What is a polymer?
- How does the molecular structure of a polymer affect its properties?
- What are some everyday items that are made of polymers?
- What properties of a polymer would make it suitable for use in a water bottle? As fibers for clothing? As a component of a motorcycle helmet?
- What polymers are found naturally in living things?

Preparing to Investigate

Polymers are long molecular chains composed of smaller repeating units called **monomers**. Natural polymers such as proteins, DNA, starch and cellulose are important to all living systems. Many common household items you use everyday are made of synthetic polymers such as polystyrene and nylon. Although the polymers that we will study are all composed primarily of carbon, their properties vary widely. We can make polymers that are opaque or transparent, rigid or flexible, weak or strong, sticky or smooth. In this investigation, you will examine the properties of several polymers, including some that are commercially available and others that you will synthesize. You also will relate the properties of the polymers to their molecular structures.

The properties of a polymer depend upon the nature of its monomer, and the overall chemical structure of the polymer, including chain length. In addition, the arrangement of the polymer chains relative to each other plays a large role in the material properties. In most polymers, the molecules are arranged randomly, as shown in Figure 25.1A. The mechanical strength of such polymers will be equal in all directions. In other polymers, however, the molecular chains are arranged in parallel to each other, as shown in Figure 25.1B. The strength of a polymer with parallel chains varies with direction. Pulling lengthwise on this polymer meets with resistance from the covalent bonds within a chain. Pulling crosswise, however, has less resistance and simply moves one chain further away from another.

Figure 25.1. A Random polymer chains and B Parallel polymer chains.

In some polymers, one molecular chain is only weakly attracted to its neighboring chains. Since the chains can easily slide past one another, these polymers often are flexible and have low melting points, Rigid chains and stronger attractions between chains result in solid polymers that are stiff or crystalline, and liquid polymers that are viscous (syrupy). In **crosslinked polymers**, the chains are covalently linked together. These polymers are non-crystalline polymers but have high viscosity and mechanical strength.

Today, producers and consumers are becoming increasingly concerned about the toxicity of polymers in household products and about the fate of polymers in recycling centers or in the landfill. Chemists are studying ways to improve the properties of polymers while making the polymers and their manufacturing processes less risky for human health and the environment. This **green chemistry** approach has led to the development of many renewable and degradable polymers that have recently come on the market.

In this investigation, you will observe and record properties of polymers. There are many noteworthy properties that you might want to consider. Some examples include:

- color
- physical state (liquid or solid?)
- viscosity (how syrupy?)
- brittleness (would it break if hit with a hard object?)
- elasticity (will it stretch and snap back?)
- adhesion (sticky or smooth?)
- tensile strength (how hard can you pull before it breaks?)
- chemical reactivity (how do the properties change upon reaction with another substance?)

Making Predictions

After reading *Gathering Evidence*, prepare a table for each test that you will perform. In the table, write the name and draw the chemical structure of each polymer to be tested, and then predict the results of the chemical or physical test that you will perform. Use what you know about bond strength and intermolecular interactions to form a hypothesis about how each polymer will perform in each test. Leave plenty of space to record your observations during your investigation. Also, draw the balanced chemical reaction for each polymer that you will synthesize.

Gathering Evidence

Overview of the Investigation

1. Observe properties of Teflon, HDPE and Mater-Bi before and after stretching.
2. Observe degradation of Teflon, HDPE and Mater-Bi with acid.
3. Synthesize and observe the properties of nylon, polyurethane foam, and/or crosslinking polyvinyl alcohol gel.
4. Measure the effects of adding different amounts of sodium borate on the viscosity of crosslinked polyvinyl alcohol gel.

Part I. Testing the Strength and Degradation of Polymer Films

In this part of the investigation, you will examine the properties of three commonly available, polymer films.

A. Tensile strength

1. **Teflon**. Obtain a piece of polytetrafluoroethylene (PTFE, or Teflon) tape (about 5 cm in length). Observe the properties of this polymer (Hint: Revisit the *Preparing to Investigate* section for suggested properties to consider.) Test the strength of the polymer by gently pulling first along its length and then pulling side-to-side. Record your observations.

2. **HDPE**. Obtain a piece of high-density polyethylene film (roughly 5 cm × 10 cm) that has been cut from a grocery-store bag. HDPE contains linear molecules of polyethylene, and has different properties from polyethylene made from branched molecules, known as low-density polyethylene (LDPE).

   a. Observe and record the properties of this polymer. Test the strength of the polymer by gently pulling first lengthwise and then side-to-side. Record your observations.

   b. Now grasp the film by its shorter sides. Pull firmly and slowly until the middle of the strip has stretched significantly. Observe and record the properties of this stretched section. Compare the width of the stretched section to that of the unstretched end portions. Test the strength of the stretched portion both lengthwise and crosswise. Record your observations.

3. **Mater-Bi**. Obtain a piece of Mater-Bi film that has been cut from a BioBag. Mater-Bi is a natural and renewable polymer formed by crosslinking amylose, a linear form of starch, to varying degrees to obtain polymers with different properties. The piece you have been given is from a bag used to collect food waste for community composing projects. Test the properties of the Mater-Bi as you have the previous two polymers and record your observations.

B. Degradation of polymers

Place a small piece of each of the three polymers into separate wells of a wellplate. Cover each sample with enough 3 M H_2SO_4 (sulfuric acid) to cover the polymer. Let the polymers sit in the acid for at least 45 minutes while you do the rest of the investigation. After this time, carefully remove the polymer pieces from the wells and rinse with water. Reexamine the properties of the polymers.

 STOP! Sulfuric acid is corrosive. Always wear eye protection and keep the solution off your skin.

Part II. Synthesizing Nylon

Nylon is a very strong polyamide fiber described in detail in Section 9.7 of *Chemistry in Context*. To prepare nylon, you will do a condensation reaction of a diacyl chloride with a diamine as shown in the reaction below.

$$n \text{ Cl}-\overset{\overset{\text{O}}{\|}}{\text{C}}-(\text{CH}_2)_4-\overset{\overset{\text{O}}{\|}}{\text{C}}-\text{Cl} \ + \ n \text{ H}-\underset{\underset{\text{H}}{|}}{\text{N}}-(\text{CH}_2)_6-\underset{\underset{\text{H}}{|}}{\text{N}}-\text{H} \ \longrightarrow \ \text{Cl}-\left[\overset{\overset{\text{O}}{\|}}{\text{C}}-(\text{CH}_2)_4-\overset{\overset{\text{O}}{\|}}{\text{C}}-\underset{\underset{\text{H}}{|}}{\text{N}}-(\text{CH}_2)_6-\underset{\underset{\text{H}}{|}}{\text{N}}-\right]_n\text{H} \ + \ (n\text{-}1) \text{ HCl}$$

In this reaction equation, the *n* and *n-1* represent the molar equivalents of each compound. In the structure of the nylon polymer, the *n* subscript indicates the number of repeat units (shown within the brackets) in the polymer chain.

1. Place 10 mL of nylon solution A (adipoyl chloride in hexane) in a 50-mL beaker. Place 10 mL of nylon solution B (hexamethylenediamine in water) in a second 50-mL beaker,

 STOP! Wear gloves when doing this investigation, and be careful not to touch the nylon with your bare hands until it has been thoroughly rinsed with water. The monomers are hazardous and the hydrochloric acid formed as a reaction by-product can cause chemical burns.

2. Tip the beaker containing solution B at a slight angle and then *slowly* add solution A by pouring it gently down the side of the beaker. Solution A should form a separate layer on top of solution B. *Do not stir or mix.*

3. A film of nylon will form between the layers. Use tweezers to grasp the film and gently pull it up and out of the beaker. Wrap the film around a test tube and then rotate the test tube to gradually pull the rest of the filament from the beaker. If you are careful, you can get one long continuous strand of nylon. If the strand breaks, simply grab it again with tweezers and wrap it around the test tube again. Continue pulling the nylon film until there is no more solution in the beaker.

4. Rinse your nylon thoroughly under a gently stream of tap water. Slide the nylon off of the test tube. Observe and record the properties of the wet nylon.

5. Spread some of your nylon strands on a paper towel to dry. Continue with other parts of the investigation but return to observe and record the properties of the nylon after it dries.

6. Dispose of the solid nylon and any excess solutions as directed by your instructor.

Part III. Synthesis of Polyurethane Foam

A urethane is a functional group formed from the reaction of an isocyanate with an alcohol. The formation of linear polyurethane from a diisocyanate with a diol depicted below is an example of urethane formation.

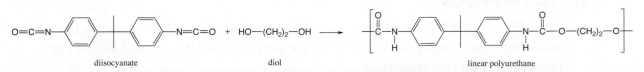

To form the **crosslinked** polyurethane foam, one or both of the monomers contain more than two functional groups, Additionally, a small amount of water is added to react with the isocyanate and form carbon dioxide gas. This reaction causes the polyurethane to foam, and surfactants and other additives enhance foam formation. The chemistry is complex, but the result is a strong and rigid, but lightweight, material that is used for applications requiring thermal insulation or impact resistance. Your instructor may give you additional information about the reaction that you will be performing.

1. Place 4 mL of polyurethane solution A (which contains the alcohol monomer) in a 3-oz. paper cup. Be patient as you pour so that you get as much of the viscous liquid into the cup as possible. Add two drops of food coloring. Mix with a wooden splint or disposable applicator stick.

2. Add 4 mL of polyurethane solution B (which contains the isocyanate monomer) to the paper cup. Use the wooden splint to *thoroughly* mix the liquids. Once the liquids are well mixed, remove the mixing stick and allow the cup to sit undisturbed.

3. The reaction may take a minute or two to begin. Once it does, record your observations. You may gently touch the outside of the cup as the reaction proceeds to observe any change in temperature, but do not touch the polymer itself yet.

4. The reaction will be complete after about 5 minutes, The polymer should no longer be sticky and can be touched. Observe and record the properties of the polyurethane foam.

5. Dispose of the polyurethane foam as directed by your instructor.

Part IV. Synthesis of Polyvinyl Alcohol Gel and Viscosity Measurements

Polyvinyl alcohol (PVA) contains a long chain of carbon atoms with alcohol functional groups attached on alternating carbon atoms. When sodium borate [$NaB(OH)_4$] is added to the PVA, it reacts with the alcohol groups to make covalent bonds between polymer chains and results in the formation of a **crosslinked polymer**, as shown in the reaction below.

Polymers with more abundant or stronger crosslinks exhibit more rigidity or **viscosity**. You will compare the properties of the original polymer to samples with varying degrees of crosslinking.

A. Synthesis of PVA Gel

1. Place 50 mL of 4% PVA solution in a 100-mL beaker. Note that pure PVA is a white solid, but it has been dissolved in water for the reaction. Observe and record the properties of this polymer solution.

2. Add 5 mL of 4% sodium borate solution to the PVA solution while stirring. Continue stirring for several minutes until the mixture becomes homogenous. You may wish to use a metal spatula rather than a glass rod to stir because the mixture becomes very thick.

3. When the reaction is complete, remove the polymer from the beaker and examine it. You can touch the polymer with your hands but wash your hands afterwards. Try forming a ball with the polymer. Does it bounce? Does it remain round? Does it stretch? Observe and record the properties of your PVA gel.

4. Dispose of the PVA gel as instructed.

B. Viscosity Measurements

In this portion of the investigation, you will determine how the amount of sodium borate solution added to PVA affects the viscosity of the resulting crosslinked polymer. Your instructor will assign you a specific amount of sodium borate to add, and your results will be combined and analyzed with those from the rest of your class.

1. Place 50 mL of 4% PVA solution in a 100-mL beaker.

2. Place your assigned volume of 4% sodium borate solution (from 0 to 5 mL) in a small beaker. Add water to bring the total volume of your solution to 5 mL.

3. Mix the two solutions, and record the time when you added the sodium borate to the PVA. Stir with a metal spatula and break up any solid lumps. Periodically mix and break up lumps for 5 minutes so that you have a homogenous gel.

4. Transfer your gel to a 50-mL beaker. Use a ruler and a marker to draw two lines exactly 30 mm apart on the side of the beaker. The top line should be somewhat below the surface of the gel and the bottom line should be somewhat above the bottom of the beaker, as shown at right.

5. Hold a steel ball above your gel so that it just touches the surface. Release the ball and record the time in seconds that it takes the ball to move from the top line to the bottom line.

6. Share your data, including the volume of sodium borate used and the time for the ball to drop, with the rest of the class, and record the other data for analysis.

7. Dispose of the gel mixture as directed by your instructor.

Analyzing Evidence

Part I

1. Which polymer was the strongest? Which was the weakest?

2. How did the strength of the Teflon and stretched HDPE change with direction?

3. Were any of the polymers affected by reaction with sulfuric acid? How did the properties change? Which of the polymers seemed to be most resistant to the reaction with sulfuric acid?

Part II

4. How did the lengthwise strength of your nylon strand differ from its crosswise strength?

5. How did the properties of the wet nylon differ from those of the dry nylon?

Part III

6. How do the properties of your polyurethane foam differ from Teflon, HDPE, Mater-Bi, and nylon?

Part IV

7. Compare the properties of the PVA solution to the PVA after adding the sodium borate solution. How is the substance different?

8. How is the viscosity of the PVA affected by adding different amounts of sodium borate? Make a graph of the class data, recording the time for the steel ball to drop (in seconds) on the *y-axis* versus the volume of sodium borate used on the *x-axis*. Can you draw a straight line through your data points, or does the relationship seem to be represented by a curved line?

Interpreting Evidence

1. Do you think the polymer strands in Teflon tape are arranged randomly or in parallel? Explain your reasoning.

2. When a high-density polyethylene film is stretched, a "neck" can form. What do you think happens to the arrangement of the polymer strands in this region of the polymer? Support your answer with investigation data.

3. Nylon is often drawn into fibers. What lengthwise and crosswise strength do you think would be important for a polymer fiber to exhibit? Explain your reasoning. Did your observation of nylon's properties match this expectation?

4. When you made nylon, you were warned that the monomers can be hazardous but that the polymer, once rinsed, is relatively harmless. Speculate on the difference in hazard between the monomers and the polymer.

5. Both polyurethane foam and polyvinyl alcohol gel contain crosslinks. What are crosslinks? How do the strengths of crosslinks in these two polymers compare to each other? Back up your assertion with investigation evidence.

6. Describe the relationship between the viscosity of polyvinyl alcohol gel and the amount of added sodium borate. Explain the connection to crosslinking.

Making Claims

What can you claim about how polymer properties are affected by the molecular structure of the polymer? Compare polymers based on chemical composition, crosslinking, and other structural features.

Reflecting on the Investigation

1. Mater-Bi is one example of the new biodegradable polymers that have come on the market in the last decade. Look online for answers to these questions. Cite your sources.

    a. Identify two other commercially available biodegradable polymers.

    b. Besides their use in the collection of waste for community composting, what other applications do biodegradable polymers have? (Hint: Have you ever had dissolving stitches to close a wound?)

2. Polyurethane solutions A and B are sometimes injected directly into the walls of older homes and allowed to foam in place. Why?

3. Teflon is commonly used as a non-stick coating on cookware. A compound used in its processing, perfluorooctanoic acid (PFOA), has been a source of public controversy. Look online for answers to these questions. Be sure to cite your sources.

    a. List two concerns about PFOA.

    b. How might the key ideas in green chemistry be used to address these concerns?

Identifying Common Plastics

Asking Questions

- What do you use everyday that is made of plastic?
- What common types of plastic are present in the items that you use often?
- Which of these items are easy to recycle in your area, and which are more difficult?
- What are the six commonly recyclable plastics designated by recycling symbols 1 through 6?
- Why must plastics get sorted for recycling?
- How can the properties of the plastics be used to sort them?

Preparing to Investigate

Plastics have been an important part of our industrial society since Leo Baekeland invented the first synthetic polymer in 1907, a thermosetting resin called Bakelite that offered an alternative to cellulose resins. Today, plastics are found in everything from food and beverage containers to furniture and electronic equipment. According to the most recent data available from the U.S. Environmental Protection Agency (EPA), about 250 million tons of municipal solid waste was generated in the United States in 2011. This waste contained 32 million tons of plastics, including 14 million tons of containers and packaging. Of all these plastics, a mere 2.7 million tons, about 8%, is recycled. This rate is well below those for other materials such as newspapers (73%), aluminum cans (55%), and glass containers (34%).

 The majority of plastic waste is composed of six common polymers. The plastics industry has adopted a code for packaging materials that can be used to identify the type of plastic in a container. The idea behind the symbol is to make recycling easier by making the identification of the plastics more straightforward. The symbol on the bottom of many containers is a triangle of arrows chasing each other with a number in the middle of the triangle.

The code for these symbols is as follows:

Number	Name	Abbreviation
1	Polyethylene terephthalate	PET
2	High-density polyethylene	HDPE
3	Polyvinyl chloride	PVC
4	Low-density polyethylene	LDPE
5	Polypropylene	PP
6	Polystyrene	PS
7	Other (includes composites and mixtures)	

Compliance in labeling is voluntary, and not all plastics have an identification code symbol. Without code numbers these plastics are difficult to classify or separate by appearance.

In this investigation, you will examine the properties of six types of plastics and use simple tests to develop a classification and identification scheme. The four tests that you will perform include:

1. The relative **density** of each plastic will be measured by checking to see whether the samples float or sink in three liquids of differing densities.
2. The second test will be to melt the plastic. All six of these common plastics melt reversibly, which means that when they are cooled, they harden and may regain their original properties. If a plastic sample does not melt, it is a **thermosetting plastic**. Thermosetting plastics, such as melamine (used in high-quality plastic dinnerware), do not melt cleanly and reversibly but tend to char instead.
3. All common plastics burn (some only if held directly in the flame), but they do so with slightly different characteristics and different noxious fumes. The vapors given off from the burning plastic may have different properties depending on the plastic. The ignition test must be performed only in a **fume hood**.
4. The copper-wire test will be used to determine if a halogen, such as chlorine, is part of the polymer.

Making Predictions

After reading *Gathering Evidence*, prepare a table to record your investigation results from the four tests on known plastics. After you have devised your classification scheme, you will need to prepare another table for your results on the unknown substances. Which polymer should give a positive flame test?

Gathering Evidence

Overview of the Investigation

1. Obtain known samples of the 6 plastics and perform four tests on each sample.
2. Devise a classification scheme and a way to identify each plastic.
3. Test your classification scheme with known samples.
4. Identify unknown samples using the four tests and your classification scheme.

Part I. Density Tests

Three liquids of differing densities will be used, as shown in the following table:s

Liquid	Density (g/mL)
47.5% ethanol in water	0.94
Water	1.00
10% NaCl in water	1.08

1. Obtain 3 test tubes and label them with the identity and density of each of the three solutions. Put about 5 mL of each liquid in the test tube labeled with its density.

2. Place two narrow strips of each of the six types of plastic in a labeled vial or beaker. Cut *one* strip of each plastic into three small pieces and store in the labeled container.

3. For the first plastic sample, place one piece into each of the three test tubes containing the density test liquids. Push each piece under the liquid surface with a glass stirring rod or the end of a pencil. If the sample floats, it has a density lower than that of the liquid. If it sinks, it has a density higher than that of the liquid. Record your observations in your data table.

4. Remove the plastic samples from the test tubes with a pair of tweezers, and then test each of the other plastics one at a time. Record your observations.

Part II. Melt Test

1. Place a small sample of each plastic, one at a time, on the end of a metal spatula and hold the end of the spatula over a light blue microburner flame.

 CAUTION! Take great care when using an open flame. Long hair MUST be tied back, and loose sleeves should be avoided. Pieces of hot molten plastic can cause burns if dropped onto your skin or clothing. Additional information on using a flame burner can be found in the *Laboratory Methods* section.

2. Heat slowly and observe the plastic as it warms and finally melts. DO NOT heat the sample so strongly that the plastic catches on fire. Enter your observations in your data table.

3. Cool the sample and examine it for appearance and flexibility by bending it. Record your observations. You may use your melted and cooled plastic for the subsequent tests.

Part III. Ignition Test

 STOP! Toxic fumes are produced during the ignition test, so it is imperative that this part of the investigation be done in a fume hood. If a fume hood is not available, do not do this part of the investigation.

1. Place a microburner or Bunsen burner and a large beaker of water in a fume hood. Light the burner and adjust it to a small flame.

2. Hold one end of a small strip of plastic in a pair of tongs, forceps, or pliers and place it directly in the flame. Observe the color of the flame and its characteristics. Is a lot of smoke or visible vapor given off? Does the plastic continue to burn after it is removed from the flame? Record your observations.

3. Test the vapors given off for acidic properties by holding a piece of *wet* litmus paper in the vapors above the burning plastic. If the paper turns red, acidic fumes are being formed as the plastic burns. Record your observations.

4. Extinguish the burning plastic by dropping it into the beaker of water. Repeat the ignition test for the other plastics.

5.

Part IV. Copper-Wire Test

 STOP! Toxic fumes are produced during the copper-wire test, so this part of the investigation MUST be done in a fume hood. If a fume hood is not available, do not do this part of the investigation.

1. Push the end of a 6-inch length of copper wire into a small cork.

2. Use the cork as a handle and heat the free end of the wire in a burner flame until the flame has no green color.

3. Touch the hot copper wire to the plastic you are testing and then return the wire end to the flame. The hot wire should pick up a tiny bit of plastic. Return the wire end to the flame. When the tip of the wire is put in the flame, watch for a slight flash of luminous flame. This indicates that you have correctly picked up a little bit of plastic on the wire.

4. Watch for the appearance of a green flame or green color in the flame when the plastic is heated in the flame. The green color indicates the presence of a halogen, such as chlorine, in the plastic.

5. Test each plastic sample and record your observations.

Part V. A Puzzle for You to Solve

1. Work ahead to the *Analyzing Evidence* section. In that section, you will analyze your results from the four tests and use this data to devise a method for identifying the six plastics that you were given.

2. After completing the work in *Analyzing Evidence*, use your classification scheme to identify at least two plastic samples that you brought from home or that your instructor has provided for you.

3. As a third step in the investigation, two unknown plastic samples will be given to you to identify. Use your scheme to determine their identities.

Analyzing Evidence

1. Rank your six plastics from least dense to most dense.

2. What differences did you observe between the plastics in the melt test?

3. What differences did you observe in the ignition test? Did one plastic stand out? If so, describe this result.

4. What differences did you observe in the copper-wire test? Did you correctly predict which plastic would give a positive test?

5. Devise a scheme for identifying the six plastics based on the simple tests you have performed. The challenge is to find the minimum number of tests that will correctly identify

the plastics when you are given any one of them as an unknown. Draw a flow chart or otherwise outline your identification scheme, then return to Part V of *Gathering Evidence*.

Interpreting Evidence

1. Were all four of the tests necessary for identifying the six plastics, or could you have focused on fewer tests?

2. Suppose you had to add two other plastics to your scheme, polymethylmethacrylate (density 1.18-1.20 g/mL) and poly-4-methyl-1-pentene (density 0.83 g/mL). Where would they fit into your scheme?

Making Claims

What can you claim about how properties of plastics can be used to identify them? Did you devise a reasonable method for identifying the different plastics?

Reflecting on the Investigation

1. Why are plastic recyclers very concerned about identifying the different polymers and not mixing them together?

2. Polyethylene terephthalate (PET) is the most valuable waste plastic at the present time. Suggest a way to separate it on a large commercial scale from other waste plastics.

3. Since waste plastic consists mostly of hydrocarbon compounds, it has been suggested that waste plastic could be used as fuel. Based on your observations in this investigation, do you think this is a reasonable suggestion? Would some plastics be more dangerous to burn than others? Defend your answer.

Notes

Identifying Analgesic Drugs with TLC

Asking Questions

- What is an analgesic?
- What compounds are often in an analgesic tablet?
- What substances other than the active ingredients are in an analgesic tablet?
- What properties of a molecule can allow it to be separated from other molecules?
- How can you identify the compounds present in a tablet?

Preparing to Investigate

Over-the-counter analgesic preparations are widely sold and used worldwide. Aspirin (acetylsalicylic acid) is a common analgesic that is used for pain and fever relief. Some people take a low dose of aspirin as a preventative medicine for heart attacks and strokes due to its anti-clotting activity. Other common non-prescription analgesic drugs include acetaminophen (also known as paracetamol), ibuprofen, and naproxen. Caffeine is sometimes added to analgesic tablets to overcome drowsiness and increase the pain-relieving properties of the medicine. Buffered aspirin contains a base such as magnesium hydroxide or calcium carbonate to neutralize the acidic compound and help prevent stomach irritation. Analgesic tablets also contain binders, such as starch and sugars, and smooth coatings to ease swallowing.

In this investigation, you will use thin-layer chromatography (TLC) to identify the analgesic compound(s) present in an over-the-counter painkiller preparation. TLC is described in detail in the *Laboratory Methods* section of this lab manual. TLC is one of the easiest chromatographic techniques and allows you to separate and identify compounds based on their differing affinity for the stationary phase (usually polar silica or alumina coated on a plastic or aluminum plate) and the mobile phase (less polar organic solvents).

Making Predictions

After reading *Gathering Evidence* and the section on TLC in *Laboratory Methods*, explain how TLC will be used to identify the components of an analgesic tablet. Look up the structures of the three most common analgesics – aspirin, acetaminophen, and ibuprofen – and caffeine, and identify structural differences that may contribute to different activity in TLC.

Gathering Evidence

Overview of the Investigation

1. Prepare a TLC developing chamber.

2. Spot known and unknown samples on a chromatographic plate.
3. Develop the chromatogram.
4. Calculate R_f values.
5. Identify components in an unknown sample.

Part I. Preparing the TLC Developing Chamber

1. Obtain a wide-mouth jar with a lid or a large beaker (400-600 mL) with a piece of plastic wrap or foil to cover it.

2. Add the solvent mixture (containing 25 parts ethyl acetate, 1 part ethanol, and 1 part acetic acid) until you have a 5-mm layer of solvent on the bottom of the container. To provide an atmosphere saturated with solvent inside the container, place a piece of filter paper around the inside surface of the container that extends into the solvent. Cover the container and set it aside while preparing the chromatographic plate.

Part II. Preparing the Sample

1. Obtain an unknown analgesic tablet (or a portion of one) and record the sample number.

2. Use a mortar and pestle to crush the tablet to a fine powder.

3. Add 2-3 mL of methanol to the powder and stir. Allow the mixture to settle.

Part III. Preparing the Chromatographic Plate

1. Obtain your chromatographic plate. Be careful not to bend or twist the sheet, and handle it only by the edges.

2. Use a pencil (not a pen!) to draw a *very light* line across the sheet in the short dimension about 1 cm from one end (see *Figure 27.1*). Make 5 small, light marks at 1-cm intervals along the line. These are the points where you will spot your samples. Label each point with a penciled number or abbreviation. Your instructor may suggest different spacing based on the size of the TLC plate and the number of spots.

Figure 27.1 Spotting the TLC plate

3. Before spotting your plate, you should practice putting *very small* spots of solution on a piece of scrap paper. To do this, dip the tip of a glass capillary tubes that have been drawn down to a very small opening into a solution. Very gently touch the tip to the paper for a brief moment. Be careful because the tubes are both sharp and fragile.

4. Use capillary tubes to spot your samples onto the TLC plate. The spots should be as small as possible in order to minimize tailing and overlapping when the chromatographic sheet

is developed. If a more intense spot is desired, let the spot dry and re-spot in the same location.

5. Known solutions of aspirin, acetaminophen, ibuprofen and caffeine will be available in the lab with a glass capillary in each. Use the capillary tubes to place small spots of the four solutions at their designated pencil marks. Your instructor may ask you to add another spot with a mixture of all four compounds. Finally use a clean glass capillary tube to spot a sample of the clear solution from your unknown tablet onto the chromatographic plate. Allow the solvent to evaporate.

Part IV. Developing the TLC Plate

1. When the spots are dry, place the plate in the developing chamber. Check to be sure that the bottom edge (near the spots) is in the solvent, but that the spots are above the solvent. Also, be certain that the filter paper does not touch the TLC plate. Cover your chamber and watch carefully. The liquid will slowly move up the TLC sheet by capillary action.

2. When the front edge of the liquid has moved to 1 cm below the top of the plate, remove the plate from the TLC developing chamber. *Immediately*, while the sheet is still wet, draw a pencil line on the sheet to show the top edge of the liquid. Then, lay the sheet on a clean surface in a fume hood or other well-ventilated area and allow the solvent to evaporate until the sheet appears dry.

3. The spots are unlikely to be visible to the naked eye, but they should be easy to see when viewed under an ultraviolet (UV) lamp. Observe your plate under the UV lamp, and carefully outline any visible spots in pencil.

CAUTION! UV radiation is harmful to your eyes. Never look directly into the UV lamp.

Analyzing Evidence

1. Calculate the R_f for your known samples and for each spot in your unknown sample. See the *Laboratory Methods* section if you need a reminder for how to do this. Report your results in a table.

2. Tape your TLC plate to a piece of paper and cover the whole plate with clear tape to protect it. Alternatively, draw your plate and dispose of it, as directed by your instructor.

Interpreting Evidence

1. Did the three pure analgesic compounds and caffeine each produce a single spot? Did the spots move different distances?

2. If you were given a known mixture of analgesic compounds to analyze, did the number of spots match the expected number of compounds? Did each spot move the same distance up the plate as the corresponding pure compound?

3. How many spots appear on your TLC plate for your unknown sample? How many active compounds are present in your unknown sample? Can you identify the compounds based on their R_f values?

Making Claims

What can you claim about the identity of your unknown sample?

Reflecting on the Investigation

1. Why do you think it was important to use a very small amount of sample when spotting the plate?

2. Suggest possible advantages and disadvantages of using a longer (taller) TLC plate.

3. If two components have an identical R_f value, does this mean they necessarily have the same structure? Explain why or why not.

4. Do you expect that changing the solvent will change the R_f value for a given component? Explain your reasoning.

5. The relative movement of components is controlled partially by the polarity of the molecules. The TLC sheet is coated with a highly polar substance, and the solvent mixture has a much lower polarity. From your chromatographic results, predict the relative polarities of aspirin, acetaminophen, ibuprofen, and caffeine. Explain your reasoning.

6. In an effort to identify an unknown, some students obtained the TLC plate shown here.

    a. How many compounds are in Sample A? In Sample B? In the unknown?

    b. A student concludes that the unknown is the same as Sample B because of the number of spots. Is this a valid conclusion? Explain.

    c. Another student concludes that the unknown is the same as Sample A because they both have spots with R_f values of 0.3. Is this a valid conclusion? Explain.

    d. Propose a reasonable conclusion for the identity of the unknown. Explain your reasoning.

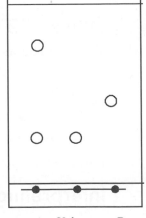

A   Unknown   B

Synthesizing Aspirin

Asking Questions

- How are useful molecules like aspirin synthesized?
- Why is it necessary to purify molecules after a chemical reaction?
- What properties can you use to separate desired molecules from impurities?
- What are some methods for determining the purity of synthesized molecules?

Preparing to Investigate

Many of the chemical substances that you use each day do not occur in nature, or if they do they do not occur in large enough amounts to meet demand. Therefore, chemists have devised ways of synthesizing and purifying useful molecules for consumer products. This investigation is a representative example of a chemical synthesis. Aspirin, a common analgesic drug is prepared from salicylic acid by a reaction known as **esterification**. You may explore two methods for synthesizing aspirin: (1) a conventional method very similar to the original synthesis of aspirin carried out by Felix Hoffman in 1893 where the reaction is heated on a hotplate and (2) a similar reaction where microwave irradiation is used for heating.

Hoffman's method for producing aspirin involved reacting salicylic acid with acetic acid in the presence of sulfuric acid as a catalyst.

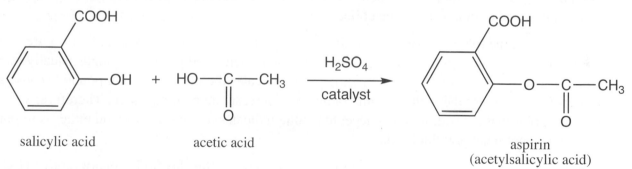

salicylic acid   acetic acid   aspirin
                                                            (acetylsalicylic acid)

For the conventional synthesis you will use acetic anhydride instead of acetic acid. Acetic anhydride is more reactive than acetic acid and is prepared by the **dehydration** of two acetic acid molecules.

acetic acid   acetic acid   acetic anhydride

Acetic anhydride works better than acetic acid in the traditional aspirin synthesis and requires less catalyst.

The reaction using acetic anhydride is shown here:

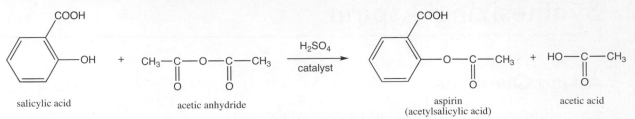

salicylic acid   +   acetic anhydride   →   aspirin (acetylsalicylic acid)   +   acetic acid

You will also use acetic anhydride in the microwave synthesis, but, in this reaction, the sulfuric acid catalyst is not needed. Microwaves are a form of electromagnetic radiation with frequencies in the range of 300 to 300,000 MHz. Household microwaves are used to cook food, but microwave sources can also be used in the laboratory to heat reactions. Microwaves interact with polar molecules and ions and cause them to rotate rapidly and heat up. Typically, reactions conducted in the microwave take place more quickly and with reduced consumption of energy compared to reactions heated in a conventional way. Thus, microwave chemistry is often considered to be a green chemistry technique.

After the aspirin is synthesized, it must be isolated and purified. The impurities in the crude aspirin include unreacted starting materials, catalyst, and byproducts. Fortunately, most of these compounds are very soluble in water and are washed away from the mixture during the isolation of the crude (unpurified) aspirin. However, salicylic acid and aspirin are only sparingly soluble in water. Therefore, any unreacted salicylic acid may be a contaminant in the crude product mixture. A technique known as **recrystallization** will be used to separate the aspirin from the unreacted salicylic acid. This technique involves dissolving the crude mixture in a suitable warm solvent. The solution is cooled to a point at which one of the components of the mixture crystallizes out of solution while the other component (the more soluble one) remains dissolved. The crystallized component can be collected by filtration.

You can examine the purity of your products by measuring the melting points. Every pure compound has a characteristic temperature at which it melts, and pure compounds usually melt sharply within a range of 1°C. The melting point indicates the purity of a compound because any impurities will lower the melting temperature and increase the melting range. Therefore, a melting point that is lower than the accepted value indicates that the compound either is impure or that it is not what you think it is.

Another way to assess the purity of your compound is using thin-layer chromatography (TLC), described in detail in the *Laboratory Methods* section and in Investigation 27. TLC can allow you to see if compounds other than the desired one are present in your sample.

Note: This investigation requires more equipment and more steps than most of the laboratory exercises in this manual. It is important for you to read the instructions carefully and be sure you understand them before proceeding.

Making Predictions

After reading *Gathering Evidence,* prepare a table to record your data, including the mass of starting materials and products and the melting points of your isolated products.

Gathering Evidence

Overview of the Investigation

1. Collect the necessary equipment and chemicals.
2. Combine salicylic acid, acetic anhydride, and sulfuric acid, and heat in a water bath.
3. Combine salicylic acid and acetic anhydride, and heat in a microwave.
4. Collect the crude products by vacuum filtration.
5. Purify the products by recrystallization.
6. Assess the purity of your products by melting point and/or TLC.

 STOP! Eye protection is absolutely essential to keep the corrosive acids out of your eyes during this investigation. If you spill acid on yourself, wash it off with large amounts of water. Keep the part of you that came into contact with acid under running water for at least 5 minutes, and notify your laboratory instructor as soon as possible. If you spill acid on your workspace, notify your lab instructor who will clean up the spill properly. If you are allergic to aspirin, notify your lab instructor before proceeding.

Part I. Synthesizing Aspirin

Materials required

- 2.1 g salicylic acid
- 3 mL acetic anhydride
- 10 drops concentrated sulfuric acid
- 2 mL distilled water
- 25 mL distilled water
- 2 250-mL beakers
- 125-mL Erlenmeyer flask
- 50-mL Erlenmeyer flask
- 10-mL graduated cylinder
- Glass stirring rod
- Ring stand and clamp
- Crushed ice and beaker for ice bath
- Small Büchner or Hirsch funnel with filter paper
- Filter flask, tubing and filtering gasket

A. Reaction Set-up

1. Fill a 250-mL beaker half-full with hot water and put it on a hot plate in the fume hood. Turn the hot plate on and bring the water to a boil. This will serve as your water bath.

2. Weigh 2.1 g of salicylic acid (0.015 mole) into a *dry* 125-mL Erlenmeyer flask. *It is crucial that the flask be dry because any traces of water will interfere with the synthesis!* Record the appearance of the salicylic acid.

 CAUTION! Sulfuric acid and acetic anhydride are corrosive, and acetic anhydride has irritating fumes. Perform the reaction in a fume hood.

3. In the fume hood, carefully measure 3 mL of acetic anhydride into a *dry* 10-mL graduated cylinder.

4. Add the acetic anhydride to the salicylic acid. Swirl to mix the liquid with the solid.

5. Carefully add 3 drops of sulfuric acid to the mixture.

6. Clamp the flask on a ring stand and lower it into the hot water bath (which should be at or close to boiling temperature). Heat for 10 minutes. If necessary, stir the mixture with a glass rod to break up any lumps. Be sure you have heated for at least 5 minutes past the point where most of the solid has dissolved.

B. Isolation of Crude Aspirin

1. Remove your flask from the hot water. Add 2 mL of distilled water, a few drops at a time, with continuous stirring. *Do not add the water all at once!* The water reacts with any remaining acetic anhydride. Solid crystals of aspirin may start to form at this point.

2. Prepare an ice-water bath by adding ice and water to a 250-mL beaker. Use approximately 2 parts ice to 1 part water. An optimal ice-water bath is mostly ice but with enough water to produce a slurry of ice and water.

3. Add 10 g of crushed ice to your flask and swirl. Cool your flask in the ice-water bath for 5-10 minutes with occasional swirling or stirring.

4. Place about 25 mL of distilled water in a 50-mL Erlenmeyer flask and cool this in the ice bath. You will use this cold water to rinse the aspirin after you filter it.

5. While your reaction mixture is cooling, prepare a vacuum filtration setup like the one shown in the *Laboratory Methods* section. Your equipment may look slightly different, and your instructor will demonstrate how to use it.

6. Attach the filtration setup to the aspirator. Insert a small Büchner or Hirsch funnel into the filter flask and put a piece of filter paper in the funnel. The filter paper should lay flat and just cover all the holes in the bottom of the funnel. Wet the paper with a few drops of water.

7. Turn the water aspirator on. There should be enough suction so that the filter paper is held firmly into the funnel. If not, seek assistance from your instructor.

8. Quickly pour the cooled slurry of product into the center of the filter paper. If your setup is working correctly, the water should be sucked through the funnel, leaving a white pasty product on the filter paper.

9. Rinse the reaction flask several times with 2-3 mL of *ice-cold* water and pour the rinses through the funnel.

10. Keep the suction on for several minutes so that air is drawn through the solid to dry it. Spread the aspirin evenly on the filter paper to help speed the drying, and pull air through the solid until it is nearly dry.

11. When the product is dry, detach the tubing from the aspirator and then shut off the water. *Do NOT shut off the water first!*

12. Transfer the solid to a pre-weighed watch glass by inverting the funnel over the watch glass and tapping gently. Reweigh and record the mass. You should have about 2 g of product.

Part II. Microwave Synthesis of Aspirin[1]

Materials required

- 0.7 g salicylic acid
- 1.4 mL acetic anhydride
- 25 mL distilled water
- 100-mL graduated cylinder
- 50-mL beaker
- 50-mL Erlenmeyer flask
- Glass stirring rod
- Microwave oven inside a fume hood
- Ice bath from Part I
- Filtration apparatus from Part I with a clean funnel and new filter paper

A. Reaction set-up

1. Weigh out 0.7 g of salicylic acid using a laboratory balance and place it into a 50-mL beaker.

2. Measure 1.4 mL acetic anhydride using a *dry* graduated cylinder and add that to the salicylic acid in the beaker.

3. Stir the mixture and place it in the microwave oven. Heat the reaction mixture at 80% power for two minutes. Remove the reaction mixture and stir it gently with the glass rod.

 CAUTION! The microwave should be placed in a fume hood, and the sash on the fume hood should be pulled down while the microwave is operating. Use caution (and perhaps a pair of gloves) when removing your beaker from the microwave since it may be very hot to the touch.

4. Heat the reaction for another 2 minutes at 80% power. Stir it gently and then allow it to sit undisturbed on the counter in the fume hood. Crystals should begin forming after about 10 minutes. If no crystals form within 20 minutes, gently scratch the inside of the glass beaker with a glass rod. If crystals still do not form, consult your instructor.

B. Isolation of Crude Aspirin

1. Once your reaction has cooled to room temperature, place the beaker in the ice bath to cool it further. Be sure to not let your beaker tip over in the ice bath or you will lose your

---

[1] Adapted from: Montes, I., *et al.* A Greener Approach to Aspirin Synthesis Using Microwave Irradiation. *J. Chem. Ed.* **2006**, *83*, 628.

product. While your reaction is cooling, place 25 mL of distilled water into a 50-mL Erlenmeyer flask, and place the flask into the ice bath to cool the water.

2. Once the reaction has completely cooled, add the cold water to the crystals in the beaker and stir. This will dissolve any unreacted acetic anhydride but will not dissolve the crystallized aspirin.

3. Filter the aspirin crystals as described in Part I above, using a clean filter and a new piece of filter paper. You should end up with about 0.8 g of crude product.

Part III. Purifying the Aspirin by Recrystallization

<u>Materials required</u>

- Crude reaction mixture from Part I
- Hot water bath and ice bath from Part I
- 50-mL Erlenmeyer flask
- 10-mL graduated cylinder
- 6 mL ethanol
- 10 mL distilled water
- Filtration apparatus with a clean filter and new filter paper

At this point, the aspirin you have obtained is somewhat wet and probably contaminated with some salicylic acid. You now will purify the crude mixture obtained from the conventional synthesis by a technique known as recrystallization. The recovered crystals from the microwave synthesis should be fairly pure and will be analyzed directly in Part IV. If you do decide to recrystallize them, be sure to reduce the amounts of ethanol and distilled water proportionally to the amount of crude product you have.

1. Save a small amount of your crude aspirin for later comparison. Place the rest of it into a 50-mL Erlenmeyer flask. Measure 6 mL of ethanol in a graduated cylinder and add it to the flask. Clamp the flask to a ring stand and heat in a hot-water bath in a fume hood. The solid should all dissolve.

2. Add 10 mL distilled water to the flask and heat until the solid dissolves again.

3. As soon as the solid has dissolved completely, take your flask back to your workstation and let it cool slowly to near room temperature. Do *not* cool rapidly with ice water. Aspirin crystals should form during cooling. If not, consult your instructor for assistance.

4. Once the flask has cooled to near room temperature and crystals are visible, cool the flask for 5 more minutes in an ice-water bath. Do not let the flask tip over or you will lose your product!

5. Collect the product by vacuum filtration as described above.

6. Spread your product out to dry on a watch glass or paper towel.

At this point, your instructor will tell you if you can leave the product to dry thoroughly until the next laboratory period, and complete the investigation then. Having a completely dry product will provide you with more accurate weights and melting points. If you complete the investigation in one day, allow your product to dry while you clean up your equipment before doing the melting point measurements.

7. Observe and describe your purified product and compare it to the crude product. Record your observations.

8. Determine the mass of your purified product by weighing it in a preweighed vial or beaker.

Part IV. Analyzing Product Purity

One of the classic ways to identify and determine the purity of an organic chemical compound such as aspirin is to measure its melting point. Your laboratory may have a special apparatus for determining the melting point of organic compounds, or an alternate procedure may be used. Your instructor will demonstrate how to obtain this measurement.

Measure the melting temperatures of salicylic acid, pure aspirin (if available), and your crude and recrystallized aspirin synthesized by conventional heating, and the product of the microwave reaction. Record the melting temperatures as ranges, recording the temperature at which the compound first begins to melt and the temperature at which the compound is completely melted.

Another way to assess the purity of your aspirin is to analyze it using thin layer chromatography, as described in Investigation 25. Your instructor may provide you with the necessary materials to compare your synthesized aspirin to standard solutions of aspirin and salicylic acid. Be sure to analyze both the crude and purified products. Alternatively, if you perform this investigation prior to Investigation 25, you may be able to save your product to analyze alongside the analgesic tablets.

Analyzing Evidence

1. How does the appearance of your recrystallized aspirin product compare to the crude product? To salicylic acid? To the product of the microwave reaction?

2. You started the conventional synthesis with 0.015 mole of salicylic acid. Do the following calculations and show your work:

    a. If the reaction worked perfectly, how many moles of aspirin would you make? (Look at the balanced equation for the synthesis reaction.)

    b. Convert this answer into grams of aspirin. To do this, multiply the number of moles of aspirin by the molar mass of aspirin (180 g/mole).

    c. Compare your actual mass of purified aspirin to the predicted mass above. What is your percent yield?

    $$\text{percent yield} = \frac{\text{actual mass of product}}{\text{predicted mass of product}} \times 100\%$$

3. You began the microwave synthesis with 0.005 moles of salicylic acid. Repeat the yield calculations described above for this synthesis. Show your work.

Interpreting Evidence

1. Draw the structures of salicylic acid and aspirin. Circle and identify the functional groups present in each compound.

2. The equation for the synthesis reaction is given in the *Preparing to Investigate* section. Copy this equation and indicate on the structures which bonds are broken in the reaction and how the resulting broken pieces combine to give the products.

3. Sulfuric acid is used as a catalyst for the conventional reaction. Define the term "catalyst". This catalyst is not added to the microwave reaction, but the reaction is still acid-catalyzed. What is the source of the acid?

4. A percent yield of 100% implies that all of your starting material was converted to product. Suggest two reasons why percent yield might be lower than 100%, and two reasons why it might be higher than 100%. Which of these applies to your reactions today?

5. Impurities tend to lower the melting temperature of a solid and make it melt over a wider range of temperatures. The published melting temperatures are 159°C for salicylic acid and 135°C for aspirin. Discuss your melting point measurements compared to these standard values. How pure are your products?

6. What impurities are present in your crude aspirin product? Where do those impurities go in the recrystallization process?

Making Claims

What can you claim about your aspirin synthesis? How efficient was your synthesis, and how pure was your product? Use evidence from your investigation to support your claims.

Reflecting on the Investigation

1. The aspirin synthesized in this investigation should *never* be taken home for medicinal use. Why not?

2. The conventional reaction will not take place if a catalyst is not added to the reaction mixture. Yet, good yields can be obtained in the microwave without a catalyst. Speculate why.

3. What safety, environmental, and economic considerations would be important to a company manufacturing 1 million kilograms of aspirin each year by the conventional procedure? Which of these considerations do you think are most important? Why? Look at the key ideas in green chemistry in the inside cover of this book. Which principles does the conventional synthesis follow well? Which does the microwave synthesis follow? Which synthesis do you think is greener? Provide support for your argument.

Drugs in the Environment

Asking Questions

- List some of the pharmaceutical and personal care products that you use each day.
- How might these products enter the environment?
- What are some consequences when these chemicals enter the environment?

Preparing to Investigate

Pharmaceuticals and personal care products (PPCPs) enhance or improve our quality of life and can enable us to live longer and more comfortably. However, due to improper disposal, incomplete metabolism, and other routes, PPCPs often end up in our waterways and soils (Figure 29.1). These compounds can have an adverse effect on the environment. Some compounds decompose slowly or do not decompose under normal environmental conditions. Others produce decomposition byproducts that are even more hazardous than the original compound. Understanding the fate of PPCPs in the environment provides important information about the proper use and disposal of these products and is an important step toward remediating the adverse effects of these materials.

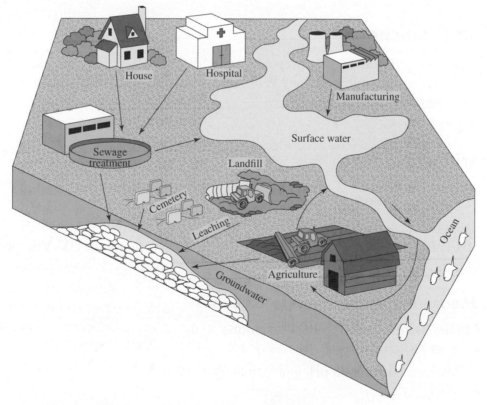

Figure 29.1. Some ways PPCPs can enter the environment.

In this investigation, you will explore the fate of caffeine under different conditions by measuring how fast it decomposes under normal and extreme environmental conditions. Here, caffeine will serve as a model drug, one that demonstrates the activity of other PPCPs found in the environment. Caffeine is appropriate for this study because it is one of the most commonly used drugs and often is found in the environment due to human use and disposal.

In chemistry, it is extremely important to know how long a reaction will take. For example, a drug molecule that takes 10 years to break down in the environment is different than a drug that takes 30 seconds. Chemical kinetics tell us about the **rate of reaction**, which is defined as how fast or slow a chemical reaction occurs. The rate of reaction is a measure of how the concentration of a reactant or product changes over time. In this investigation, you will study the rate of reaction for the degradation of caffeine under various pH conditions.

In order to determine a reaction rate, it is necessary to directly monitor the concentration of a reactant or product over time. For this investigation we will be using **UV-Vis spectroscopy**. Some molecules absorb ultraviolet (UV) or visible (Vis) light. The amount of light absorbed is related to the concentration of that molecule in solution, as shown in Equation 1.

$$\text{Absorbance (A)} = b \times \text{Concentration (C)} \tag{1}$$

In this equation, absorbance (A) is the amount of light absorbed by a sample, and b is an experimental constant. This equation shows us that the absorbance of light is correlated with concentration. Measuring the change in absorbance over time will allow you to determine the rate of reaction.

Making Predictions

After reading *Gathering Evidence*, prepare a table for recording your data. How do you think pH will affect the degradation of caffeine? Will the degradation reaction be faster or slower in acidic, basic, or neutral solutions?

Gathering Evidence

Overview of the Investigation

1. Use a pH meter to measure the pH of the reaction solutions.
2. Mix caffeine with each solution and monitor the reaction with UV-Vis spectroscopy.
3. Plot the data to determine reaction rates.

Part I. Measuring pH of Reaction Solutions

1. Obtain 20 mL of each reaction solution.
   - 1 M HCl
   - Acetate buffer with pH approximately equal to 5
   - 9 M NaOH

CAUTION! Hydrochloric acid is corrosive and can cause burns. Avoid contact with skin, and always wear safety glasses.

2. Carefully calibrate your pH meter. Operation of the pH meter is described in the *Laboratory Methods* section of this lab manual.

3. Measure and record the pH of your three reaction solutions. Thoroughly rinse the electrode between each measurement, and pat it dry using a soft tissue.

Part II. Measuring Decomposition Rate of Caffeine

This part of the investigation is time sensitive and must be done quickly but carefully. Make sure that you have all equipment ready to use before mixing the caffeine with the reaction solution. Work with one reaction solution at a time, and carefully clean and dry your glassware before moving on to the next reaction.

1. Familiarize yourself with the operation of the UV-Vis spectrometer that you will be using. Set the measurement wavelength at 272 nm, and perform a blank measurement using a cuvette containing 1 M HCl.

2. Using a volumetric pipet, measure 3 mL of the caffeine solution into a 50-mL beaker, and then pipet 3 mL of the 1 M HCl reaction solution into a second 50-mL beaker.

3. Quickly but carefully, pour the contents of one beaker into the other and mix for three seconds. Then, transfer the mixture into a quartz cuvette so that the cuvette is ¾ full.

4. Place the cuvette into the sample holder of the spectrometer and immediately measure the absorbance of the sample. **NOTE:** If done correctly, this step should take no more than 10-15 seconds. You must record your initial absorbance reading as soon as possible after mixing the reagents.

5. Obtain absorbance measurements every 15 seconds for 5 minutes. Record the time after mixing and the absorbance for each measurement in your data table.

6. Dispose of your solutions in appropriate waste containers. Wash and dry the beakers and cuvette, then repeat the mixing and measurement with the acetate buffer.

7. Clean everything again and repeat the measurement process with 9 M NaOH.

Clean-up

Dispose of all solutions in an appropriate waste container. Wash glassware with soap and water and allow it to drain.

Analyzing Evidence

1. Using a computer graphing program, create a plot of absorbance vs. time for each of the three reaction mixtures. Time should appear on the x-axis of the graph, and absorbance should be plotted on the y-axis. Which reagent(s) caused a decrease in caffeine concentration over time?

2. For any reagents that did cause a change in concentration of caffeine, determine an average rate of decomposition between the initial measurement and 1 minute later. To do this, use the graphing program to calculate the slope of the decomposition line between these two times. Then, determine the average rate between 4 and 5 minutes. Are these rates the same?

Interpreting Evidence

1. For any reaction that showed a decomposition of caffeine, how does the rate of decomposition change with time? Does the reaction speed up or slow down as it proceeds? Explain your results.

2. The acetate buffer was chosen to mimic the natural pH of ground water throughout the country. Use the U.S. EPA website to look up the pH of the groundwater in your area. Does the buffer adequately represent the conditions near you? Do you think caffeine that finds its way into your local environment will decompose quickly or slowly?

Making Claims

What can you claim about how pH affects the decomposition of caffeine?

Reflecting on the Investigation

1. Other investigations in this lab manual have asked you to measure the pH of water containing dissolved CO_2, and your textbook discusses this, too. Is carbonated water acidic or basic? Based on your answer and the results of this investigation, do you expect the caffeine in carbonated beverages to decompose or be stable?

2. Look up and draw the structure of caffeine, and identify the functional groups.

3. The human digestive tract is acidic (the stomach contains approximately 0.16 M HCl). Under these conditions will caffeine decompose quickly or slowly? Do you think that the caffeine you consume in your morning coffee will completely break down in your body, or will some be excreted later in the day? Once you formulate a hypothesis, search for evidence to back it up. Cite your sources.

4. What are some consequences of caffeine contamination in the environment?

5. Will all drug molecules exhibit the same reactivity as caffeine? Why or why not?

6. Do a search for another PPCP that is found in the environment due to human activities. The U.S. EPA website can give you some ideas. Write a short paragraph explaining how the molecule enters the environment and the consequences of its presence to wildlife and the ecosystem as a whole.

This investigation was adapted from an experiment developed by Michael Samide and Todd A. Hopkins at Butler University: *J. Chem. Ed.* **2013**, *90*, 1162-1166.

Measuring Fat in Potato Chips and Hot Dogs

Asking Questions

- How much fat is in potato chips and hot dogs?
- What physical property can we use to separate the fat from foods in order to measure how much is present?
- What are some health consequences of excess fat in the diet?
- Do different types of chips or different types of meat products contain different amounts of fat?

Preparing to Investigate

Fats are an essential part of our diet, but consuming too much fat has been linked to health problems including obesity and heart disease. In this investigation, you will perform an **extraction** to measure the amounts of fat in foods that are a common part of our modern American diet. Extractions work by separating compounds in a mixture (such as a food) by their solubility. Two procedures will be used to determine the fat content of foods. In potato chips, the fat is mostly on the outside, so the fat can simply be dissolved using a suitable solvent. However, since the fat is bound up in the protein within meat, another approach must be used for hot dogs and other meat products. These samples must be chemically treated to break down the protein in and liberate the fat. The fat will then float on top of water and can be skimmed off the surface easily.

Making Predictions

After reading *Gathering Evidence* make a data table for your results. Develop several scientific questions that you can answer with this investigation, and predict the relative fat content of the samples that your class will analyze.

Gathering Evidence

Overview of the Investigation

1. Weigh the food samples.
2. Grind up the chip samples and extract the fat using petroleum ether.
3. Evaporate the petroleum ether and weigh the fat residue.
4. Mix the meat samples with protein-liquefying reagent.
5. Centrifuge the meat samples, and then separate and weigh the fat.
6. Calculate the percentage of fat in your samples.

Part I. Measuring Fat in Potato Chips

The procedure for extracting fat from potato chips is very simple: Petroleum ether (a commercial mixture of hydrocarbons that is widely used as a solvent) is mixed with ground-up chips to extract the fat. After separating the solvent mixture from the chips, the petroleum ether is evaporated and leaves the fat that can be weighed. This procedure will work well for potato chips, tortilla chips, crackers, or any other similar type of snack. You may wish to investigate the differences in fat content between different brands or types of chips or assess whether products that claim to be "low fat" actually fit this description.

1. Obtain clean, dry beakers for your samples. You will need one beaker for each sample. Label the beakers with names or abbreviations for your samples.

2. Weigh the beakers and record the mass of each beaker.

3. Weigh 2-3 grams of each sample you will analyze. Record the mass to the nearest 0.01 g.

4. Use a clean porcelain mortar and pestle to crush your first sample into small pieces.

5. Add 15 mL of petroleum ether to the mortar and grind the mixture thoroughly.

 CAUTION! Petroleum ether is a gasoline-like solvent that is *extremely* flammable. It is absolutely essential that no open flames be present anywhere in the laboratory during this investigation. Under no circumstances should the heating be done with a Bunsen burner.

6. Separate the petroleum ether from the solid chip residue by performing a gravity filtration. This procedure is further explained in the *Laboratory Methods* section. Prepare a glass funnel with folded filter paper and mount it over the appropriate beaker. Use a spatula to transfer the potato chip mixture into the filter.

7. Rinse the mortar twice, using 5 mL of petroleum ether for each rinse, and put the liquid into the filter to drain into the beaker.

8. Repeat the procedure for each sample. Use a clean, dry mortar and pestle and a new piece of filter paper for each sample.

9. When all samples have been extracted, you will remove the petroleum ether by evaporation. There are two ways this can be done:

    a. You can leave the beakers in a fume hood until the next day or the next lab period, and the solvent will evaporate.

    b. A much faster way is to place the beakers on a steam bath or hot plate *in a fume hood*. Leave them for about 15 minutes or until all of the petroleum ether has evaporated. Dry the outside of the beakers and let them cool.

10. Reweigh the beakers and record the masses to the nearest 0.01 g.

Part II. Measuring Fat in a Hot Dog or Other Meat Sample

This procedure can be used for any meat sample, which allows you to compare different samples. You may wish to investigate whether different brands of hot dogs differ significantly in fat content, whether chicken or turkey hot dogs differ from those made with pork or beef, whether hot dogs and hamburgers differ significantly in fat, whether "low fat" products are really as claimed, or how vegetarian hot dogs compare to meat hot dogs in their fat content.

1. Put 75 mL of water in a 250-mL beaker. Place the beaker on a hot plate and heat the water to between 80° and 90° C.

2. While the water is heating, label a centrifuge tube for each sample that you plan to analyze. Weigh and record the mass of each tube to the nearest 0.01 g.

3. Put between 2 and 2.5 g of ground meat in each test tube and weigh the tubes again.

4. Add about 5 mL of the protein-liquefying reagent to each meat sample. The protein liquefying reagent is a solution of sodium salicylate, potassium sulfite and sodium hydroxide in isopropyl alcohol and water.

 CAUTION! The protein-liquefying solution is highly caustic and can cause serious skin damage. Do not allow it to contact your skin! If you do get it on yourself, immediately wash your skin with copious amounts of water and notify your instructor.

5. Put the centrifuge tubes in the beaker of hot water and heat until the reagent in the tube starts to boil (about 80° C). Maintain the temperature of the water so that the contents of the centrifuge tubes boil for 10 minutes. *Do not leave the beaker and tubes unattended.* The reaction must be monitored to ensure that liquid does not boil out of the test tube. In addition, vapor from the protein-liquefying reagent is extremely flammable so care must be taken.

 STOP! Never use a Bunsen burner to heat a flammable substance. Always use a hot plate.

6. After the mixture has boiled for 10 minutes, it should be dark brown with some yellow fat floating on the top. Remove the centrifuge tubes from the hot water, stand them in an empty beaker, and let them cool until you can handle them but the fat is still liquid. Do not let them cool to room temperature because if the fat solidifies the separation will become difficult.

7. Obtain small containers (vials, beakers, test tubes, or watch glasses) for each of your samples. Label, weigh, and record the mass of each container.

8. Put your sample tubes in a centrifuge and make sure they are balanced. Centrifuge the mixtures to separate the fat from the rest of the solution. It should be a layer on top of the solution.

9. Use a Pasteur pipet with a rubber bulb to carefully transfer the fat from the top of your sample into the appropriately labeled container. Be very careful to remove all of the fat but none of the brown liquid. Work slowly; it takes patience to do this correctly so that only the fat is removed. Repeat this for each of your samples.

10. Weigh the containers of fat and record their masses.

Clean-up

Discard the filter paper and chip mixtures from Part I in a designated container. The "brown protein liquid" from Part II and your isolated fats should be deposited in appropriate waste containers. Wash the mortar, pestle, beakers and tubes thoroughly and allow them to drain.

Analyzing Evidence

In order to compare your samples, you should calculate the percent fat in each one. First, calculate the mass of sample (chips or meat) and the mass of fat that you isolated by subtracting the mass of the container from the mass of the container with sample or fat. Then, calculate the percent using the following equation:

$$\% \text{ fat} = \frac{\text{mass of fat}}{\text{mass of sample}} \times 100\%$$

Interpreting Evidence

1. Your instructor may ask you to share your results with your classmates. How closely do your results agree with those of your classmates for the same samples? Can you suggest any sources of error in the measurements?

2. Looking at the class results, answer the scientific questions you posed in *Making Predictions*.

3. Can you draw any generalizations about snack chips? For example, do potato chips have more or less fat than tortilla chips?

4. If your class analyzed different kinds of hot dogs or other meats, what can you conclude about the fat content in these products? Are some products significantly lower in fat than others?

5. If your class analyzed any "low-fat" or "no-fat" products, do the results support this claim? If not, suggest a reasonable explanation.

6. A simple calculation will allow you to determine if your results agree with the fat content reported by the chip manufacturer. The mass of a potato chip is mostly fat, carbohydrate, protein and water. If we assume the water content is low, dividing the reported fat content by the sum of masses of the three main components will provide you with an estimated percent fat content of the chips.

$$\% \text{ fat from package} = \frac{\text{fat content (in grams)}}{\text{sum of fat} + \text{carbohydrate} + \text{protein (in grams)}} \times 100\%$$

If the package for your potato chip sample is available, perform this calculation. Do your investigation results agree with this calculation? If not, why not? Was the assumption about water valid?

Making Claims

What can you claim about the fat content of snack chips and meat products? Can you draw any conclusions about the relative fat content of different kinds of foods?

Reflecting on the Investigation

1. Potatoes are naturally low in fat, so why do potato chips contain lots of fat? How are low-fat chips different?

2. Why did hot dogs require a different fat extraction procedure from the one used with chips? Which of the two procedures do you think would work best when analyzing roasted peanuts, corn, or pepperoni pizza? Explain your reasoning for each food.

3. A snack pack of potato chips or other kinds of chips holds 1 ounce (28 grams) of chips. Based on your data, how much fat is present in a snack pack? In such a bag of chips, there are about 15 grams of carbohydrate and about 1 gram of protein. Given that carbohydrates and protein provide about 4 Calories of energy per gram, and fats provide about 9 Calories per gram, what is the total energy equivalent (in Calories) of a 1-ounce bag of chips? What percent of the Calories are from fat? Show your calculations.

4. A typical hot dog weighs about 45 grams. Use your data to calculate the number of grams of fat in the hot dog you analyzed. There are about 7 grams of protein in a hot dog, and almost all of the remaining mass is water. Using the Calorie values given in Question 3, calculate how many Calories are provided by the fat, the protein, and the water. What percentage of the Calories in the hot dog can be attributed to the fat?

Notes

Measuring the Sugar Content of Beverages

Asking Questions

- How does the amount of sugar dissolved in a beverage affect its physical properties?
- Can we make a device to measure the amount of sugar in beverages?
- How much sugar is in soft drinks?
- How does the amount of sugar in fruit juice compare to that in soft drinks?

Preparing to Investigate

Many of the beverages we consume each day, including fruit juices and non-milk-based soft drinks, contain sugar. Have you ever wondered just how much sugar is present in these beverages? In this investigation, you will build a simple device called a **hydrometer** that you will use to measure and compare the sugar content of a number of common beverages.

How can we measure the amount of sugar in a solution? Rather than directly measuring the concentration of sugar, which would require specialized equipment, we will instead measure the **density** of the solution, a physical property related to the amount of sugar. The density is the mass (in grams) per volume (in cubic centimeters or millimeters) of a substance. For instance, the density of water is 1.000 g/mL (or g/cm^3).

When sugar (or another solute) dissolves in a liquid such as water, the volume of the solution does not increase significantly over that of the pure liquid. However, the mass of the solution increases. How will this affect the density of the solution?

Making Predictions

After reading *Gathering Evidence*, prepare a data table for your results. Develop several scientific questions that you can answer in this investigation.

Gathering Evidence

Overview of the Investigation

1. Prepare a hydrometer using a thin-stem plastic pipette.
2. Calibrate your hydrometer using sugar solutions of known concentration.
3. Use your data to prepare a calibration curve showing how hydrometer depth changes with sugar concentration.
4. Use your hydrometer to measure the density of several beverages.
5. Use your graph to determine the sugar concentration in your samples.

Part I. Constructing a hydrometer

You will construct a hydrometer as shown in Figure 31.1.

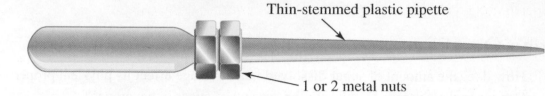

Thin-stemmed plastic pipette

1 or 2 metal nuts

Figure 31.1. A simple hydrometer.

1. Collect a thin-stemmed plastic pipette and one or two metal nuts of the type to fit on machine screws.
2. Slide the nut(s) over the stem of the pipette as shown in the figure.
3. Fill the pipette approximately half-full of water.
4. Place the pipette, bulb end down, into a 50-mL graduated cylinder filled nearly to the top with water at room temperature. Adjust the amount of water in the pipette so that it floats with the bulb *near* the bottom (but not touching the bottom) and with only a short length of the stem sticking out of the water, as in Figure 31.2. If it floats too high or too low, adjust the height by cautiously adding or removing some water from the bulb of the pipette. When you have the pipette floating correctly, it is ready to use as a hydrometer.

To determine the relative density of a liquid, and hence the amount of dissolved solids (mostly sugar) in soft drinks or fruit juices, all you need to do is float the hydrometer in the beverage and measure the length of the stem that sticks out of the liquid. Before proceeding, discuss with your partner or lab group what you expect to happen when the hydrometer is placed into a beverage that contains sugar. Will more or less of the stem protrude from the surface of the liquid?

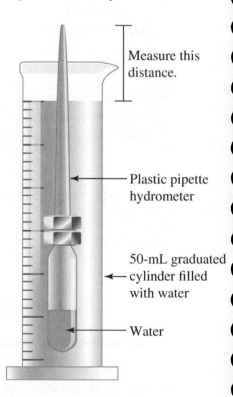

Measure this distance.

Plastic pipette hydrometer

50-mL graduated cylinder filled with water

Water

Part II. Calibrating the hydrometer

The hydrometer you have constructed cannot measure the absolute concentration of sugar in a solution, but only the relative amount. In order to quantitatively estimate the amount of sugar in various beverages, you will need to calibrate the hydrometer you have constructed. In the lab, you will have beakers, graduated cylinders, stirrers, and sugar cubes available to prepare reference solutions, and rulers to measure the length of pipette stem protruding from each solution. In your lab group, come up with a procedure for calibrating your hydrometer. How will you prepare the reference solutions? How many solutions should you make? How many sugar cubes will you put in each solution?

Figure 31.2. The hydrometer assembly

Have your instructor check your procedure before doing the calibration. Record your results in a table, and then plot the numbers on a graph (either by hand or on a computer). What should the axes of the graph be? Can your data points be connected by a straight line? Do you need to repeat any of the measurements? What sources of error might there be?

Part III. Testing beverages

Using your calibrated hydrometer and the beverages provided by the instructor in the lab, set about answering the questions you developed in the *Making Predictions* section. Devise an investigation plan that will enable you to make claims about the sugar content of the beverages with respect to your questions.

Make certain that soft drinks have been decarbonated, or follow any given instructions for doing this. If the hydrometer is placed in a carbonated drink, bubbles sticking to the hydrometer will add buoyancy and the reading will be incorrect.

To avoid contaminating the samples, always rinse the hydrometer with pure water and blot it gently with a tissue before placing it in a new solution. Use the graph you generated in part 2 to relate the height of your hydrometer stem to the concentration of sugar in the beverage. Record your data in a table, and share your results with the rest of your class. How do the results on the same beverage compare?

 STOP! Do not drink any beverages that have been opened or used in the laboratory.

Optional: If your instructor permits, you may take your hydrometer with you (being careful not to squeeze out any of the water) and use it to test other drinks. It is very important that you not drink any samples in which the hydrometer has been immersed. Measurements should be done on small samples and then discarded.

Analyzing Evidence

Use your calibration curve to determine the sugar content for each of the beverages you measured. Use your results to answer the questions you developed in *Making Predictions*.

Interpreting Evidence

1. Which beverages have the highest sugar content, and which the lowest? Did any of your results surprise you?

2. Do the results for any "sugar-free" beverages tested support this claim?

3. How does the sugar content of fruit juice compare to that of soft drinks?

4. Describe how you could modify your hydrometer to increase its accuracy.

Making Claims

What can you claim about the sugar content of beverages?

Reflecting on the Investigation

1. When cans of regular and diet soda are placed in a large container of ice water, some of the cans float while others sink. Which float? Which sink? Explain.

2. Ethanol (an alcohol) is found in beer and wine. Look up the density of ethanol, and speculate how the density of a mixture of water and ethanol will change with increasing concentration of ethanol. Devise a method to determine the concentration of ethanol in alcoholic beverages using a hydrometer. How might you modify your hydrometer to make it suitable for measuring ethanol content of alcoholic beverages?

3. Ethylene glycol is a commonly-used automobile radiator coolant. Most automobile coolant systems contain ethylene glycol mixed with water. Look up the density of ethylene glycol, and devise a procedure for measuring the concentration of ethylene glycol in an automobile cooling system.

4. Find a bottle or can of regular soda and look at the label. What is the serving size? Is it the same as the size of the container? How many carbohydrates (in g) are there per serving? How many in the whole container? Given that one teaspoon of sugar (one sugar cube) has a mass of approximately 4 g, how many teaspoons of sugar would you consume if you drank the entire container of soda?

5. Discuss the environmental and health benefits of following a diet that contains foods low in added sugar.

6. Explain how the sugar present in fruit juice differs in chemical structure and origin from the sugar you used to make your reference solution, and from that found in soft drinks and colas. How do you think this affects your data?

7. In the United States, added sugar in processed foods is mostly in the form of high fructose corn syrup (HFCS), not sucrose. There has been some controversy over the use of HFCS in foods, and many labels now claim that products are free of HFCS. Look into both sides of the issue, and discuss the environmental and health impacts of HFCS vs. sucrose. Do you feel comfortable consuming foods containing HFCS? What about sucrose?

Measuring Salt in Food

Asking Questions

- Why is salt added to food?
- What limits are recommended for salt consumption by people?
- Why should we be aware of consuming too much salt?
- Which foods contain the most salt?

Preparing to Investigate

A moderate amount of salt (sodium chloride, NaCl) is essential in our diet for good health, but consumption of large quantities can have many adverse health effects. Large quantities of salt are commonly added to many processed foods and often are hidden or masked by other flavoring agents. In this investigation, you will measure the salt content in typical servings of some packaged soups and in the liquid from a jar of pickles.

The easiest way to measure salt content in foods is to measure the concentration of chloride ions by means of a **titration**. The chemistry of the chloride ion (Cl^-) titration is described in detail in Investigation 17 and will be reviewed here briefly. A solution of silver nitrate ($AgNO_3$) is added to a measured quantity of liquid sample until just enough has been added to react exactly with all of the chloride ions present in the sample. The reaction forms silver chloride (AgCl), an insoluble white solid.

$$AgNO_3(aq) + Cl^-(aq) \rightarrow AgCl(s) + NO_3^-(aq)$$

If the amount of silver added is known, then the chloride ion concentration can be calculated because we know from the reaction equation that 1 mole of Ag^+ reacts with 1 mole of Cl^-.

In order to better see the endpoint of the titration, you will add an **indicator**, sodium chromate (Na_2CrO_4). Chromate ion (CrO_4^{2-}) reacts with Ag^+ to form silver chromate (Ag_2CrO_4), an insoluble red precipitate. As you add silver nitrate to your sample, silver ions will react with the chloride ions to form AgCl. Once all of the chloride ions have reacted, the additional silver ions will react with chromate ions to form the red precipitate. The presence of the red color signals the end of the titration.

For this investigation you will assume that all of the chloride that you measure comes from sodium chloride, NaCl. This assumption is not always a good one because some foods may have chloride ions present that are associated with other metal ions such as calcium, potassium, or magnesium. However, for processed or preserved foods such as bouillon cubes, canned soup or pickles, the majority of the chloride ions do indeed come from added salt.

Making Predictions

After reading *Gathering Evidence,* prepare a data table. Read the labels on your assigned samples, and rank them according to salt content.

Gathering Evidence

Overview of the Investigation

1. Devise a procedure for titrating chloride in foods.
2. Perform the titration on dissolved bouillon cubes, soup, and pickle liquid.
3. Calculate the salt concentrations in the samples.

 CAUTION! Be careful not to spill the silver nitrate or sodium chromate solutions on your skin. Silver nitrate will stain your skin black, and, while this is harmless and wears off in a few days, it is unattractive while it lasts. More seriously, sodium chromate is a **carcinogen,** and contact must be avoided.

Part I. Developing Your Procedure

The general technique of titration is described in the *Laboratory Methods* section. The procedure for the titration of chloride with silver nitrate is described in detail in Investigation 17. Read the background and procedure in Investigation 17, and adapt that procedure for use with foods. Write down your procedure and have your instructor check it before proceeding.

Part II. Performing Titrations on Food Samples

A. Bouillon and other soups

1. Obtain a bouillon cube and dissolve it in 1 cup, or 227 mL, of water. The cube will dissolve faster if you crush it, heat the water, and stir. Be sure to mix the solution until the cube is completely dissolved. Note that one bouillon cube will provide enough broth for several student groups, so you can share your sample.
2. Following the procedure you developed in Part I, do the titration starting with 20 drops of broth. Record the number of drops of silver nitrate required to reach the endpoint of the titration. Do at least three titrations on the broth. If you are uncertain about the accuracy of any of your titrations, perform one or two more.
3. If other soups are available, do at least three titrations on those samples. Record the number of drops used. Rinse your sample pipet thoroughly before each new sample.

B. Pickle liquid

1. Pickle liquid is typically very high in salt, so you should use a smaller volume of the sample than you did for the bouillon. Start each titration by putting 5 drops of pickle liquid into each well. Add 15 drops of water to each well to dilute the pickle liquid to 1/4 of its original concentration.

2. Perform your titration at least three times with the pickle liquid. If the first titration uses too many or too few drops of silver nitrate, adjust the dilution of pickle liquid accordingly for the subsequent trials.

Clean-up

Empty your solutions into the designated waste container, and clean your equipment thoroughly.

Analyzing Evidence

A detailed explanation of the required calculations is given in Investigation 17. After reading the appropriate section in that investigation, perform the required calculations using your data from this investigation.

1. First, look carefully at your titration results for each sample. Do the results all agree with each other reasonably well? If one of the results seems out of line compared to the others, it is reasonable to omit that result. Simply draw a line through that row of your data table (do not scribble it out!) and write a note saying why you are omitting that result.

2. Calculate the average number of drops of silver nitrate solution that were required to titrate each sample, and record this value on your data table.

3. You can calculate the molarity of the chloride ion in the sample using Equation 1 below, which was derived in Investigation 17. Do this calculation and record the molarity of chloride for each sample you analyzed.

$$\text{molarity of chloride} = \text{molarity of AgNO}_3 \times \frac{\text{drops of AgNO}_3}{\text{drops of water sample}} \qquad \textbf{(1)}$$

4. We aren't really interested in the *molarity* of chloride ions in the soups or pickles, but how much salt and, more particularly, how much *sodium* is present in a serving. Because the formula for salt is NaCl, the sodium ion (Na^+) molarity is equal to the chloride ion (Cl^-) molarity as in Equation 2.

$$\text{molarity of Cl}^- = \text{molarity of Na}^+ \qquad \textbf{(2)}$$

The mass of one mole of sodium is 23 grams or 23,000 milligrams. Thus, multiplying the molarity of sodium ions by the molar mass gives you the mass, in milligrams, of sodium per 1 L of solution. For example, suppose you measure the concentration of chloride to be 0.20 M. The sodium concentration will also be 0.20 M, so

$$\frac{0.20 \text{ moles sodium}}{1 \text{ L of soup}} \times \frac{23,000 \text{ mg sodium}}{1 \text{ mole sodium}} = \frac{4600 \text{ mg sodium}}{1 \text{ L of soup}}$$

Multiplying that value by the size of 1 serving provides the amount of sodium per serving. Supposing a serving size of 1 cup, or 227 mL, and given that 1 L = 1000 mL,

$$\frac{4600 \text{ mg sodium}}{1000 \text{ mL soup}} \times \frac{227 \text{ mL soup}}{1 \text{ serving of soup}} = 1044 \text{ mg sodium per serving}$$

Perform this calculation for all servings of soup that you analyzed. For the pickle liquid, calculate the milligrams of sodium in 100 mL of liquid.

Interpreting Evidence

1. Look up the recommended daily intake for sodium. If you consumed 1 cup (227 mL) of the bouillon or other soup that you analyzed, what percentage of the recommended daily intake of salt would you have consumed?

2. If a pickle is 90% fluid, how much sodium would be found in a 90 gram pickle of the type you estimated? What percentage of your recommended daily intake of salt would be supplied by one pickle?

Making Claims

What can you claim about salt in foods?

Reflecting on the Investigation

1. Design an investigation procedure to determine how much salt is contained in a small bag of potato chips or in a small bag of salted peanuts. In designing the investigation consider the following:

    a. How would you choose a representative sample of the chips or nuts?

    b. How would you remove the salt from the food sample?

    c. If you washed the salt off the chips or nuts, would it matter how much water you used?

    d. With solid samples, you may wish to use different units than you used for liquid samples. How would you report your results?

2. Packaged lunchmeats also contain large amounts of salt. How could you estimate the amount of salt in these products?

Measuring Vitamin C in Juice and Tablets

Asking Questions

- How much vitamin C is in juice and vitamin tablets?
- How can we measure the amount of vitamin C?
- Why is vitamin C important to our diet?
- What are macronutrients and micronutrients? Which of these terms applies to vitamin C?

Preparing to Investigate

Vitamin C, or ascorbic acid ($C_6H_8O_6$) is one of the essential vitamins required for good health. It is chemically similar to the simple sugar glucose ($C_6H_{12}O_6$), which is plentiful in our bodies. Most animals possess an enzyme needed for making ascorbic acid from glucose, but humans and a few other species lack that enzyme. Therefore, we must secure ascorbic acid directly from foods that we consume. The best-known function of ascorbic acid is the prevention of scurvy; a minute amount (10 mg per day) is adequate for this purpose. However, it also plays several other important roles in human health. Most of its functions are related to the fact that it has a strong tendency to transfer electrons to other chemical substances and is therefore classified as a reducing agent or antioxidant. It acts to prevent other, potentially harmful, reactions that involve the transfer of electrons.

To maintain good health, the current recommended daily intake (RDI) for vitamin C is 60 mg. Foods like fruits and vegetables with a high water content often contain large amounts of vitamin C. Ascorbic acid is very soluble in water so, if high doses are ingested, much of it is excreted rapidly in urine.

In this investigation, you will analyze fruit juice and vitamin C tablets to determine how much vitamin C is contained in each. You will use the **titration method of analysis**, which is described in detail in the *Laboratory Methods* section of this lab manual. The chemistry of the vitamin C titration is based on its tendency to lose electrons. A suitable electron acceptor, in this case iodine (I_2), is used as the reagent to titrate the ascorbic acid. The chemical reaction is

$$C_6H_8O_6 + I_2 \rightarrow C_6H_6O_6 + 2\ I^- + 2\ H^+$$

Since it is difficult to prepare iodine solutions with an accurately known concentration, you will first need to titrate your iodine solution using a reference sample of ascorbic acid. Because iodine is colored, the end point of the titration is signaled by the appearance of the color of unreacted iodine, which happens when all of the ascorbic acid has reacted. The color is made more intense by adding a drop of starch solution, which forms a deep blue color with iodine.

Making Predictions

After reading *Gathering Evidence*, prepare a data table in which to record your investigation results. Develop several scientific questions you can answer by analyzing the Vitamin C content of the samples available in your lab. Looking at the labels on the samples, predict the relative amounts of vitamin C in each of the samples you will titrate. Which samples should have the most vitamin C, and which should have the least?

Gathering Evidence

Overview of the Investigation

1. Prepare a solution by dissolving a vitamin C tablet in water in a volumetric flask.
2. Titrate the ascorbic acid reference solution with iodine solution.
3. Titrate the vitamin C tablet solution and any juice samples using an iodine solution of known concentration.
4. Calculate the amount of vitamin C in your samples.

Part I. Preparing the Vitamin C Tablet Solution

1. Weigh a vitamin C tablet and record its mass to the nearest milligram (0.001 g).

2. Place the tablet in a clean volumetric flask. Fill the flask about half full with water. Shake and swirl the flask until the tablet is broken down. It may be helpful to gently crush the tablet with a glass stirring rod. Label the flask and allow it to settle.

Part II. Titrating an Ascorbic Acid Reference Solution

Before proceeding, review the section on titration in the *Laboratory Methods* section at the beginning of this laboratory manual. You will be titrating a solution of ascorbic acid of known concentration in order to determine the concentration of your iodine solution. The procedure given here for the titration will be used for titrating your samples as well.

1. Use a graduated-stem plastic pipet to add exactly 1 mL of the ascorbic acid reference solution into each of four wells in a clean, dry wellplate. Place the wellplate on a white surface to make it easier to observe color changes.

2. Add 1 drop of starch solution to each well.

3. Obtain a small supply of iodine solution in a clean container. Use a pipet to add one drop of iodine at a time to your solution. Count the drops and stir between additions. You'll notice a momentary blue color that disappears with stirring but lingers longer as more iodine is added. The end point occurs when one drop of iodine gives a pale blue color that does not disappear with stirring. In your data table, record the number of drops that lead to this color change.

Part III. Titrating Vitamin C Tablet Solution

By now, the vitamin C tablet in the volumetric flask should be mostly dissolved, except for the insoluble filler material that is usually starch. Add deionized water to the flask up to the bottom of the neck. Then use a pipet to add water *one drop at a time* until the bottom of the curved meniscus of the liquid surface just touches the etched line on the neck of the flask (Figure 33.1). Cap the flask securely and mix the contents of the flask by repeatedly turning the flask upside down and swirling.

Rinse the graduated-stem pipet you used for the reference solution with deionized water (filling and emptying several times), and then rinse with solution from the volumetric flask in the same manner. Use the pipet to add exactly 1 mL of the vitamin tablet solution to each of four clean wells in the wellplate. Add a drop of starch to each well, and continue the titration as described in Part II. Record the number of drops required to titrate the solution in your data table.

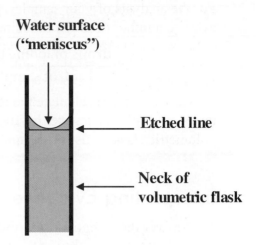

Water surface ("meniscus")

Etched line

Neck of volumetric flask

Figure 33.1. Using a volumetric flask

Part IV. Titrating Juice Samples

Use the same procedure described above to titrate one or more samples of juice. You can use bottled juice, or you can extract and analyze the liquid from fruits and vegetables that are high in vitamin C (such as peppers and broccoli). You should do at least four titrations for each sample. Be sure to thoroughly rinse your graduated-stem pipet between samples with deionized water and with the new sample to be analyzed before carefully adding 1 mL of sample to each well in the wellplate. Record your data.

 CAUTION! Do not consume any food or beverage used in the laboratory.

Analyzing Evidence

Look at your data table and decide if the data for each set of titrations shows consistency or whether any individual result should be excluded because of an error. If so, make a note beside that value and omit it from your calculations. Then, calculate the average number of drops of iodine used for each set of titrations.

You will use your data from the titration of the reference solution to calculate the *calibration factor* for this analysis. This will give you a way of correlating the amount of ascorbic acid in your samples to the amount contained in the reference solution. To calculate the calibration factor, you will need to divide the concentration of the reference solution, in mg/mL, by the number of drops of iodine used in the titration, as shown in the equation below. You will then know how many milligrams of ascorbic acid are titrated by 1 drop of iodine.

$$\frac{\text{mg ascorbic acid}}{\text{1 mL reference solution}} \times \frac{\text{1 mL reference solution}}{\text{drops of iodine used}} = \frac{\text{mg ascorbic acid}}{\text{1 drop of iodine}}$$

For the analysis of your samples, you simply need to multiply your calibration factor by the average number of drops of iodine used for each titration, as shown below.

$$\frac{\text{drops of iodine used}}{1 \text{ mL of sample}} \times \frac{\text{mg ascorbic acid}}{1 \text{ drop of iodine}} = \frac{\text{mg ascorbic acid}}{1 \text{ mL of sample}}$$

This allows you to calculate the milligrams of ascorbic acid present in 1 mL of sample. For the tablet, multiply the answer by 100 to find the total milligrams of ascorbic acid in the 100-mL volumetric flask. This is the same as the milligrams of ascorbic acid in one tablet.

Interpreting Evidence

1. Answer the scientific questions you posed in the *Making Predictions* section.

2. What fraction of the mass of the vitamin tablet was ascorbic acid?

3. How closely did your analysis of the vitamin tablet match the label on the bottle? Calculate the percentage difference between them.

4. If a standard serving of juice is 8 ounces and 1 fluid ounce = 30 mL, calculate how much vitamin C would be obtained from an 8-ounce serving of the juice you analyzed. Does this match the amount cited on the label? How many servings would you need to consume to meet the RDI for vitamin C?

Making Claims

What can you claim about the amount of vitamin C in various foods and juices?

Reflecting on the Investigation

1. Is vitamin C a water-soluble or fat-soluble vitamin? How does this relate to the procedure we used for analysis?

2. The RDI cited in the introduction is the recommendation to the general population by the U.S. Food and Drug Administration for the minimum amount that will allow most people to maintain health. In 2004, the Institute of Medicine (a branch of the National Academies) published more detailed Dietary Reference Intake (DRI) values that take into account age, sex, and activity levels of individuals when recommending nutrient intake levels.
    a. Locate the tables, and find your DRI for Vitamin C. How much vitamin C should you take in each day?
    b. How much juice or other foods that you analyzed would you need to ingest in order to meet your DRI?
    c. Do you feel that the DRI is a better suggestion of nutrient intake than RDI? Why or why not?
    d. What foods do you eat each day that are rich in vitamin C? Do you think that your diet provides you with enough vitamin C to meet your DRI?

Isolating DNA from Plant and Animal Cells

Asking Questions

- What property of DNA can be used to isolate it from cells?

- Why can the lysis and precipitation process be used to extract DNA from both plant and animal cells?

- How is DNA from different types of cells the same? How is it different?

Preparing to Investigate

The cells of all living organisms contain a molecule called deoxyribonucleic acid (DNA) that provides the necessary instructions for growth and reproduction of that organism. In this investigation, you will isolate DNA from a banana, an onion, or your own cheek cells. Although the procedure you will use in this investigation is very simple, it is similar to the way that hospitals and laboratories isolate DNA for medical and forensic applications.

DNA is a long polymer consisting of repeat units called **nucleotides** (see Chapter 12 of *Chemistry in Context*). Each nucleotide is composed of a phosphate group, a deoxyribose sugar, and one of four nitrogen-containing bases – adenine, cytosine, guanine, or thymine. The bases on two complementary strands form pairs through hydrogen-bonding and twist into a double-helix shape (see Figure 12.6 in *Chemistry in Context*).

Many methods have been developed for extracting DNA from cells. These procedures are designed to extract only the DNA and leave behind other cell components including proteins, lipids, and other molecules. In general, DNA isolation methods begin with a **lysis** step, where the breakdown of cell walls is promoted by heat and reaction with a detergent, a salt, and a reagent that controls pH. Once any remaining solids are removed by filtration, the DNA is **precipitated** from solution using an organic solvent such as ethanol or isopropanol. This process allows it to be isolated from other soluble components of the cell. DNA molecules precipitate in long strands that mat together and look something like coagulated egg white. The strands can be wrapped around a thin glass or plastic rod and removed from the liquid.

Making Predictions

- After reading *Gathering Evidence*, rank the solutions containing the banana cells, onion cells, and cheek cells from highest to lowest in how much DNA will be isolated by the written procedures. Explain why you predict these results.

- Prepare a data sheet that includes space for predictions, observations, and any procedural changes.

Gathering Evidence

Overview of the Investigation

1. Prepare your sample of banana, onion, or cheek cells.
2. Homogenize and heat your sample.
3. Chill the mixture and filter it to remove the cell walls.
4. Precipitate the DNA with alcohol.
5. Spool the DNA around a glass or plastic rod.

Extracting and Isolating DNA

Follow the lysis and precipitation procedure that has been designed for your sample.

A. DNA from Banana Cells

1. Chill a bottle of isopropyl alcohol by placing it inside a bucket or other large container and surrounding it with ice.

2. Make homogenizing solution by combining 10 mL of Woolite detergent, 1/2 teaspoon of NaCl (table salt), and 90 mL of water in a blender and blending at high speed until mixed.

3. Pour the blended solution into a beaker, and place the beaker into a hot water bath (at about 60°C) for approximately 15 min. After the solution has warmed, use a hand mitt to carefully remove the beaker from the hot water bath and pour the solution back into the blender.

4. Add 1/2 of a banana to the blender and blend it at high speed with the heated soap/salt solution. Return the banana mixture to the beaker and place the beaker in an ice water bath for 5 minutes to cool.

5. Carefully place 2 pieces of cheesecloth over a funnel placed in a beaker. Filter the mixture remove any unmashed pieces of banana through the cheesecloth and collect it in the beaker.

6. Place 6 mL of banana filtrate into a test tube. Then, tilt the tube and carefully pour 9 mL of cold isopropyl alcohol down the side. Try to prevent the alcohol from mixing with the banana mixture.

7. Allow the isopropyl alcohol to sit on the banana mixture for five minutes without disturbing the test tube. Bubbles will begin to form and rise up from the bottom of the test tube to the surface of the isopropyl alcohol. The banana DNA will begin to rise and precipitate out of solution. The DNA will be cloudy white and is extremely fragile, so take care to avoid any sudden disturbance to the test tube.

8. Gently place a glass stirring rod or a toothpick into the test tube without disturbing the solution. Carefully swivel it and watch the DNA wrap around the surface.

B. DNA from Onion Cells

1. Weigh 10 grams of chopped onion into a large test tube.

2. Add 20 mL of homogenizing solution (see Step 2 in previous section) and place the test tube in a 60°C water bath.

3. Keep the onion mixture in the beaker of 60°C water for 10 minutes. Monitor the temperature and be sure that it stays close to 60°C. If the temperature of the onion mixture reaches 70°C, the DNA structure will be destroyed, and you will not be able to isolate and spool it.

4. After the onion mixture has heated for 10 minutes, put the test tube in an ice-water bath and chill until the mixture reaches 15°C.

5. Pour the cooled solution into a mortar and use the pestle to mash the onion to a smooth paste.

6. Transfer the paste and all the liquid to a 50-mL beaker and chill in an ice bath for 15 minutes.

7. Use a funnel lined with coarse filter paper to filter the mixture into a 50-mL graduated cylinder. Place the cylinder containing the filtered liquid in the ice bath for 5 minutes.

8. Slowly pour 20 mL of 95% ethanol down the side of the graduated cylinder to form a layer on top of the onion liquid. A white stringy precipitate will form at the liquid/liquid interface.

9. Slowly pour the liquid into a small Petri dish. Then put the end of a thin glass rod or a glass Pasteur pipet into the liquid and slowly twirl it around. White strands of DNA should wrap themselves around the end of the pipet.

C. DNA from Human Cheek Cells

1. Pour 10 mL of fresh tap water or bottled water into a clean, 30 mL plastic drinking cup. Put the water into your mouth and swirl for at least 30 seconds. Swirling the water for longer is better because it will wash more cells from the inside of your cheeks into the water. Spit the water back into the plastic cup.

2. Add 1 mL of 8% NaCl solution and 5 mL of the "cheek cell" water to a large test tube.

3. Add 1 mL of 10% sodium lauryl sulfate solution **or** 1 mL of 25% liquid dishwashing detergent solution to the test tube. This detergent ruptures the cell membranes to release the DNA into the salt solution.

4. Place the test tube into a 55°C water bath for 5 minutes. Heating enhances the action of the detergent and also denatures enzymes that might damage the DNA.

5. Remove the tube from the water bath, place a stopper on top, and mix the contents by gently inverting the tube several times. Do not shake the tube.

6. To a new test tube, add 1 mL of 95% ethyl or isopropyl alcohol. Place this tube in a beaker full of ice to chill it.

7. Use a clean glass stirring rod to transfer a small amount of the cheek cell solution into the test tube containing the alcohol. Observe the DNA strands floating in the alcohol.

Analyzing Evidence

1. Describe the appearance of the mixture prepared with your sample and the detergent or homogenizing solution. Be sure to include any changes you observe as the reaction occurs.

2. Describe what happens as the solution comes into contact with the ethanol or isopropanol.

3. Describe the appearance and physical properties of the DNA you isolate.

Interpreting Evidence

1. In this investigation, you were cautioned not to allow the temperature to go above 60°C. Why do you think DNA decomposes when it is heated to too high a temperature? (**Hint:** Think about the double-helix structure.)

2. DNA was precipitated out of a water solution by adding ethanol (CH_3CH_2OH) or isopropanol. Why do you think DNA is less soluble in ethanol than in water? Remember that ethanol is partly a hydrocarbon and therefore is a less polar molecule than water.

Making Claims

What can you claim about the isolation of DNA from plant and animal cells? What evidence from this investigation supports your claims?

Reflecting on the Investigation

1. Below are the structural formulas for the four nucleotide bases in DNA: adenine, thymine, cytosine, and guanine. Name three features that they have in common.

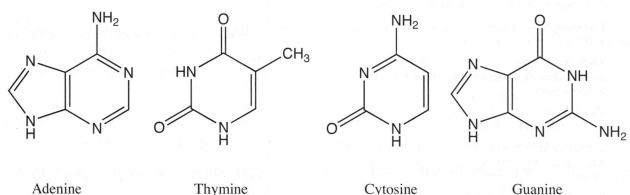

Adenine   Thymine   Cytosine   Guanine

2. Which hydrogen atoms in the 4 nucleotides are responsible for forming the hydrogen bonds that hold DNA strands together in the double helix? How many hydrogen bonds are possible for each? See *Figure 12.7* of *Chemistry in Context* for help. Suggest an explanation for why adenine always pairs with thymine and cytosine always pairs with guanine.

3. In light of your answer above, suggest why DNA molecules that are rich in cytosine and guanine will "melt" or decompose at a higher temperature than DNA that is rich in adenine and thymine.

4. The diversity of life depends on the fact that a great many combinations of the 4 nucleotide bases (shown above) can be assembled into long chains. Illustrate this by writing out 20 of the 64 possible combinations of just 3 bases, using any combinations of the four building blocks (designated as A, T, C, and G).

Glossary

Absorbance (A) – the portion of light that gets absorbed (or does not get transmitted) by a sample in spectroscopy

Acid – a chemical that donates H^+ to water.

Alkalinity – the amount or concentration of basic substances present in a solution; usually measured by titration with acid

Barometer – an instrument used to measure atmospheric pressure

Base – a chemical that accepts H^+ from water, or donates OH^- to water.

Battery – a collection of galvanic cells wired together

Biodiesel – a biofuel prepared through a transesterification of triglycerides with a small alcohol.

Biofuel – energy source derived from a biological, usually renewable, source rather than from petroleum

Blank sample – a sample containing only solvent or air, used to determine baseline absorbance or transmittance in spectroscopy

Calorimetry – a method for measuring the heat released by a substance upon combustion or other chemical reaction

Carcinogen – a substance that is capable of causing cancer

Chemical mole – the amount of a substance containing 6.02×10^{23} atoms or molecules

Chromatography – a method for separating compounds based on their differing affinities for a stationary phase and a mobile phase

Combustion – the reaction of a fuel with oxygen, that results in the release of energy

Critical point – the temperature and pressure beyond which a substance does not have distinct liquid and gas phases

Crosslinked polymers – rigid polymers in which the polymer chains are covalently linked together

Dehydration – a chemical reaction that results in the loss of water

Density – the mass of a substance divided by its volume

Distillation – a separation process in which a liquid solution is heated and the vapors are condensed and collected

Electrolytic cell – a type of electrochemical cell in which electric energy is converted into chemical energy; it can be considered the opposite of a galvanic cell, where chemical energy is converted to electric energy

Electromagnetic spectrum – the range of all possible frequencies of electromagnetic radiation

Esterification – a chemical reaction that produces an ester functional group, usually by reaction of a carboxylic acid with an alcohol

Extraction – the use of a solvent to separate soluble compounds in a mixture from insoluble compounds

Extrapolation – estimating values beyond the measured data on a graph

Fume hood – a laboratory apparatus that draws air and toxic fumes away from the chemical researcher

Galvanic cell – an electrochemical cell that converts the energy released in a spontaneous chemical reaction into electrical energy

Gravity filtration – a process to separate a liquid and a solid where the solid collects in a filter while the liquid travels through the filter unassisted except by gravity; used when a scientist wishes to collect and retain the liquid

Hydrocarbons – compounds that contain only hydrogen and carbon

Hydrophilic – a compound or substance that readily dissolves in water

Hydrophobic – a compound or substance that does not dissolve in water

Green chemistry – a philosophy that encourages the reduction of waste and the use of less toxic compounds and processes in chemical synthesis and engineering

Indicator – a chemical substance that undergoes a visible change upon reaction

Interpolation – estimating values between measured points on a graph

Lysis – breakdown of cell walls through a chemical reaction

Mass – The amount of matter in a sample

Meniscus – the curved surface of a liquid

Miscible – describes two liquids that can freely mix in any proportion

Mobile phase – in chromatography, the part of the system that moves over the stationary phase; usually a liquid or a gas

Molar mass – the mass, in grams, that contains 6.02×10^{23} (Avogadro's number) of atoms of an element, or molecules of a compound.

Molarity – the number of moles of a substance in exactly 1 liter of solution

Monomer – a small molecule used to prepare a polymer, or the repeat unit of a polymer

Nanometer (nm) – a unit of distance often used for the wavelength of visible light, equal to 1×10^{-9} meters

Natural polymer – see *polymer*

Nucleotide – repeat unit of DNA consisting of a phosphate group, deoxyribose sugar, and nitrogen-containing base

pH – a measurement of the acidity or basicity of a solution

Plastics – polymers that can be molded into a shape while warm and then set into a more rigid form

Polymers – long molecular chains composed of smaller repeating units. *Natural polymers* occur in biological systems, while *synthetic polymers* are prepared in the laboratory.

Precipitate – removal of a soluble molecule from solution by addition of a different solvent or a salt that causes the molecule to become insoluble

Pressure – the amount of force exerted by a substance on its surroundings.

Rate of reaction – how fast or slow a chemical reaction occurs, or how the concentration of a reactant or product changes over time.

Regression – a best-fit straight line for a set of data plotted on a graph

Recrystallization – a method to purify a solid compound by dissolving it in hot solvent and then cooling the solution so that pure crystals form

R_f – short for "retention factor" or "retardation factor"; in thin-layer chromatography, R_f is the distance traveled by a compound up the plate, divided by the distance traveled by the solvent.

Slope – the steepness of a linear portion of a graph, given the rise divided by the run

Solder – a low-melting mixture of several metals that is melted onto wires to make an electrical connection.

Solubility – the maximum mass of solid that can dissolve in a given volume of water

Specific heat – the amount of energy required to raise the temperature of one gram of a substance by 1° C.

Spectrophotometer – a specialized laboratory instrument that measures the absorption of light by matter.

Spectroscopy – the use of light to study matter

Spectrum – a graph depicting absorbance or % transmittance of a sample at different wavelengths of light

Stationary phase – in chromatography, the part of the system that doesn't move; usually a solid

Sublimation – a phase change in which a substance transforms from a solid to a gas

Supercritical fluid – a phase existing beyond the critical point, when a substance exhibits characteristics of both liquid and gas

Synthetic polymer – see *polymer*

Temperature – a measure of the kinetic energy of a sample

Thermometer – the instrument used to measure temperature

Thin-layer chromatography (TLC) – a type of chromatography involving a stationary phase consisting of a glass, plastic, or aluminum plate coated with solid silica and alumina, and a mobile phase consisting of a liquid solvent.

Titration – a method of analysis that uses a chemical reaction to determine the concentration of a substance; an indicator is usually used to provide a visual cue that a reaction has occurred

Total dissolved solids – the amount of dissolved material (usually ionic salts) present in a sample of water

Total hardness of water – a measure of the concentration of calcium and magnesium ions in a water sample

% Transmittance (%T) – in spectroscopy, the amount of light that passes through a sample, rather than being absorbed

Triple point – the temperature and pressure at which a substance exists in three phases: solid, liquid, and gas

Vacuum filtration – a method for separating liquids from solids whereby the solid collects in a filter while the liquid is driven through the filter by suction; used when a scientist wishes to collect and retain the solid

Viscosity – the ability of a liquid to resist the movement of a solid object through it. *Viscous* liquids are thicker and more syrupy.

Volume – the amount of space taken up by a sample. The volume of liquid samples is usually measured using a *graduated cylinder*.

Wavelength – the length of one wave of electromagnetic radiation; light of longer wavelength is lower in energy.

Wet chemistry – chemical reactions or analyses performed in the liquid phase

y-intercept – the value of a linear portion of a graph where the line crosses the y-axis, or where x=0